Thatched Huts and Stucco Palaces

Thatched Huts and Stucco Palaces

Peasants and Landlords in 19th-century Nepal

MAHESH C. REGMI

VIKAS PUBLISHING HOUSE PVT LTD
New Delhi Bombay Bangalore Calcutta Kanpur

VIKAS PUBLISHING HOUSE PVT LTD

5 ANSARI ROAD, NEW DELHI 110002
SAVOY CHAMBERS, 5 WALLACE STREET, BOMBAY 400001
10 FIRST MAIN ROAD, GANDHI NAGAR, BANGALORE 560009
8/1-B CHOWRINGHEE LANE, CALCUTTA 700016
80 CANNING ROAD, KANPUR 208004

ISBN 0 7069 0672 1

1V02R3401

PRINTED AT PROMINENT PRINTERS, K-9, NAVIN SHAHDARA, DELHI 110032

In the memory of my father

"Assemble simple and quiet farmers from every corner of the world, and they will all readily agree that they should be permitted to sell the surplus of their grain to their neighbours, and that the law to the contrary is inhuman and absurd; that the currency representing produce should no more be debased than the fruits of the earth; that the father of a family should be the master at home; that religion should bring men together in order to unite them, and not to make them into fanatics and persecutors; that those who work should not deprive themselves of the fruit of their labours to endow superstition and idleness. In an hour they would make thirty laws of this kind, all useful to mankind.

"But let Tamerlane arrive in India to subjugate it, and then you will see only arbitrary laws. One will oppress a province to enrich one of Tamerlane's tax-collectors; another will make it a crime of *lèse-majesté* to have spoken ill of the mistress of the first valet of a rajah; a third will lay hands on half the farmer's crop, and dispute his right to the rest; and finally there will be laws by which a Tartar beadle will come to seize your children in the cradle, make the most robust one into a soldier, and the weakest a eunuch, and will leave the father and mother without resource and consolation."

Voltaire: *Philosophical Dictionary*

Preface

Thatched Huts and Stucco Palaces: Peasants and Landlords in 19th-century Nepal is the first volume of a projected two-volume study of Nepal's economic history during the nineteenth century. As the title indicates, it limits itself to a study of agrarian relations. The second volume will deal with finance and trade.

This study has been undertaken in the belief that "history consists essentially in seeing the past through the eyes of the present and in the light of its problems," and that "the main work of the historian is not to record, but to evaluate."[1] It marks a departure from the general tradition of Nepali historiography, in which greater emphasis has been laid on the auxiliary sciences of history—archaeology, epigraphy, numismatics—than on history itself as "a dialogue between past and present, not between dead past and living present, but between living present and a past which the historian makes live again by establishing its continuity with the present."[2]

Inasmuch as economic development is the leading national slogan in Nepal today, this study represents an attempt to explore some of the historical and institutional constraints facing such development. In other words, it seeks to answer the question: Why is Nepal poor?

Poverty is, of course, a complex socio-economic phenomenon for which no monocausal explanation can be adequate. Nevertheless, no social scientist would deny that the elimination of poverty requires an all-round increase of productivity in the economic field. "Productivity—or output per man-hour—depends largely, though by no means entirely, on the degree to which capital is employed in production," and "is a function, in technical terms, of the capital intensity of production."[3] It follows that poverty is endemic in any agrarian society where the peasant, as the actual cultivator of land, the most important economic resource, is compelled to share the major portion of his produce with parasitic groups who have no role in production, and whose income from the land is not available for use as capital in increasing agricultural productivity.

The present volume, therefore, concentrates on relations between the state and rent-receivers on the one hand and the peasant on the other during the nineteenth century in order to identify those parasitic groups and the form and nature of their share in agricultural production. It starts with a general description of land and agriculture in nineteenth-century Nepal (Chapter 1), followed by an account of its government and politics, particularly after the commencement of Rana rule in 1846 (Chapter 2). It describes how agricultural lands were alienated on a tax-free basis by the state to individuals comprising the aristocracy and the bureaucracy (Chapter 3), and how these categories of the landowning elite collected payments both in money and in kind from the peasantry (Chapter 4). The book then discusses village elite groups that were created in order to collect rents and taxes on behalf of the landowning elite and the government (Chapter 5). This is followed by a description of the labour-tax obligations of the peasantry (Chapter 6). From these manifestations of the political domination of the peasant, the book proceeds to a broad description of the local agrarian community, in which an attempt has been made to analyze the nature and origin of property in land (Chapter 7). The economic domination of the peasant by the village moneylender forms the subject-matter of Chapter 8, and the agricultural development policies of the government, of Chapter 9. In the concluding chapter, an attempt has been made to analyze broadly the impact of the complex of legal, economic and social relations between landlord and peasant in nineteenth-century Nepal on the agricultural economy as a whole.

The book is an outgrowth of the research undertaken by the author while writing *Land Tenure and Taxation in Nepal* (Berkeley: University of California Press, 1963-68, 4 vols.), *A Study in Nepali Economic History, 1768-1846* (New Delhi: Manjusri Publishing House, 1972), and *Landownership in Nepal* (Berkeley, Los Angeles, and London: University of California Press, 1976). It represents an attempt to probe into some aspects of the economic realities that lay behind the institutional structure of agrarian relations as described in those studies. It thus marks one more step forward toward understanding in total perspective the agrarian scene of nineteenth-century Nepal. In the process, several key themes pertaining to the agrarian system have had to be reiterated and coordinated in a wider perspective. Hence their discussion in this study should by no means be considered a rehash of those earlier studies.

The study is essentially empirical in nature. It purposely seeks to avoid reference to ideal conceptual systems such as patrimonialism and bureaucracy, or feudalism and capitalism, in the belief that a conceptual analysis of society is no substitute for socio-economic research.[4] The need to avoid reference to "comprehensive and encyclopaedic social theories" is particularly important in a little-known society such as that of Nepal, situated in a part of the world where socio-economic processes have often been blanketed under such striking but incomprehensible terms as "Asiatic mode of production" and "Oriental despotism." Such simple models are clearly insufficient if justice is to be done to the very complex socio-economic forms of historical reality.

The author, of course, has no illusions that the book does any such justice; indeed, it presents only a very shadowy picture of the historical reality of nineteenth-century Nepal. The defect is inherent in the compartmental approach to the study of history. Economic history seeks to portray only one aspect of the life of our ancestors, albeit an important aspect. A proper understanding of the economic life of a nation during any particular period in its history necessitates an understanding of its political history, the history of its social institutions, including, in a country such as Nepal, of caste and communal organization, religious and cultural ideas and ideologies, literature, and so on. Nor can the choice of period be based on historical reality, inasmuch as history essentially represents a continuum. There is no alternative, nevertheless, to the compartmental approach, for human effort has a finite quality that belies a total perspective. The study will have served its purpose if the relevance of the questions it raises, the quality of the evidence it presents, and the methodology it follows make a contribution, howsoever humble, toward preparing the groundwork on which future research in this and allied fields may hopefully be based.

Any shortcomings in this study may then be balanced by the insights and illuminations that it may provide to other scholars in approaching the subject from wider or different perspectives.

The author hopes that those scholars and practical men who may read this book will value it more for the questions it stimulates than for the answers it postulates. He has framed a question the answer to which, in his view, necessitates an approach to history that has so far eluded Nepali historiography. He has thereby sought to arrive at a meaningful interpretation of nineteenth-century Nepali history that establishes its connection with the Nepal of today. The author is aware that the same evidence that he has presented may be interpreted in widely different ways to arrive at conclusions which may be equally, if not more, valid. He can, therefore, only conclude:

> In the wide ocean upon which we venture, the possible ways and directions are many; and the same studies which have served for this work might easily in other hands not only receive a wholly different treatment and application, but lead also to essentially different conclusions. Such indeed is the importance of the subject that it still calls for fresh investigation, and may be studied with advantage from the most varied points of view. Meanwhile we are content if a patient hearing is granted us, and if this book be taken and judged as a whole.[5]

MAHESH C. REGMI

NOTES

[1]E.H. Carr, *What Is History*, Middlesex, Penguin Books, 1975, p. 21.

[2]E.H. Carr, *The New Society*, London, 1951, cited in John Madge, *The Tools of Social Science*, Longmans, 1965, p. 110.

[3]Ragnar Nurkse, *Problems of Capital Formation in Under-Developed Countries*, Oxford, 1974, p. 9.

[4]Madge, *op. cit.*, pp. 69-70.

[5]J. Burckhardt, *The Civilization of the Renaissance in Italy*, trans. by S.G.C. Middlemore, Phaidon Press ed., 1944, p. 1, cited in John B. Morrall, *The Medieval Imprint*, Middlesex, Penguin Books, 1967, p. 8.

Acknowledgements

The author feels happy at this opportunity to express his gratitude to the concerned authorities of His Majesty's Government, whose gracious permission enabled him to procure copies of the archival materials used in the study; to the Nepal Studies Association, whose financial assistance (through the Center for South Asia Studies at the University of California, Berkeley) facilitated the use of these materials; to Dr Leo E. Rose, whose steady friendship, encouragement, and stimulating criticism have been of immeasurable help in the author's research efforts over the years; to Fr. Dr L.F. Stiller, S.J., but for whose friendly advice and incisive comments the book would have been considerably less readable; to Mr H.K. Kuloy and Mr Aran Schloss, who painstakingly read the initial drafts of the manuscript and offered much valuable advice; and, finally, to the Ramon Magsaysay Award Foundation of the Philippines, and, especially, its Executive Trustee, Miss Belen H. Abreu, for the 1977 Ramon Magsaysay Award for Journalism, Literature, and Creative Communication Arts, which has bolstered both his self-confidence and his credibility.

MAHESH C. REGMI

Contents

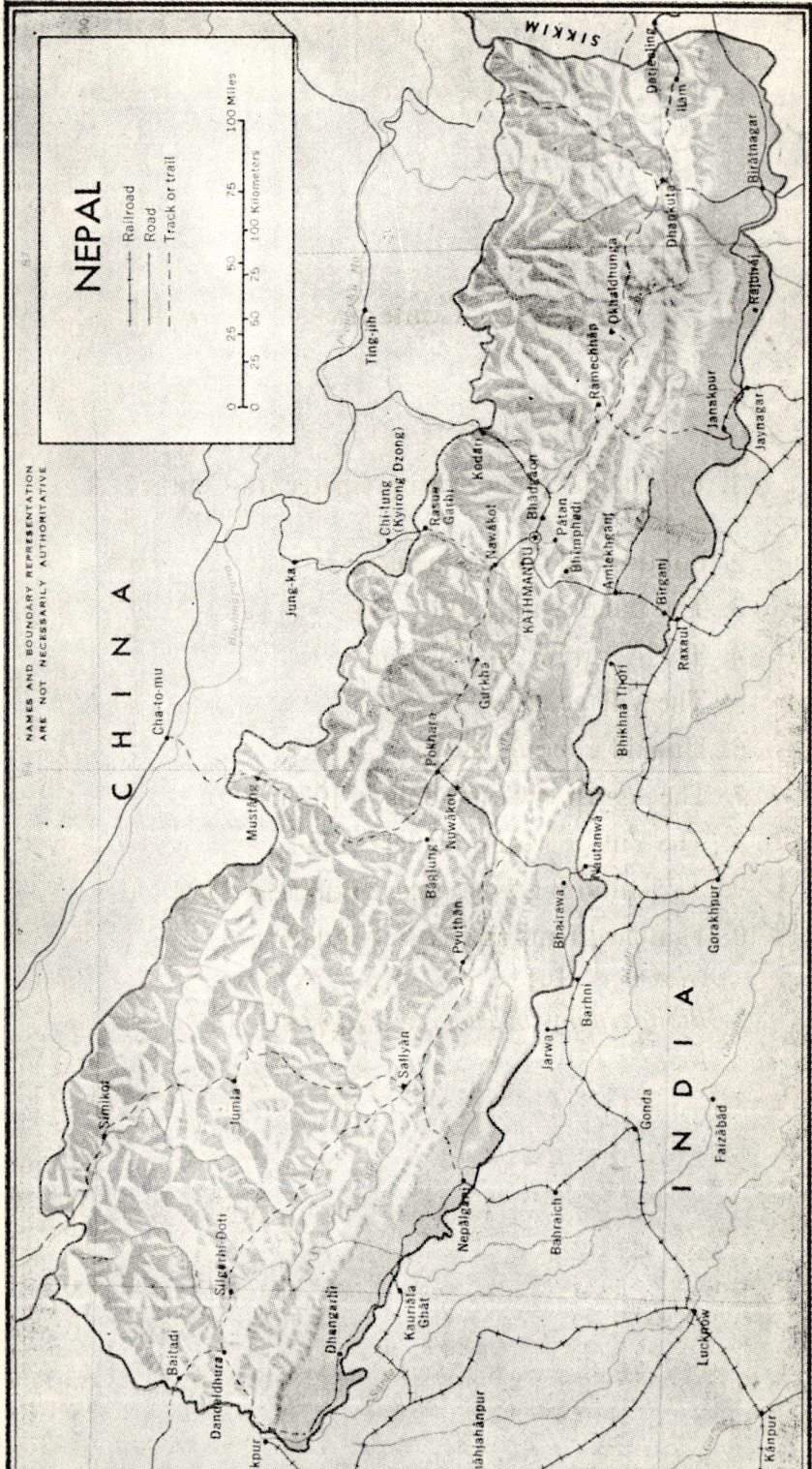

NEPAL

Railroad
Road
Track or trail

0 25 50 75 100 Miles
0 25 50 75 100 Kilometers

NAMES AND BOUNDARY REPRESENTATION
ARE NOT NECESSARILY AUTHORITATIVE

C H I N A

I N D I A

SIKKIM

Cha-to-mu

Ting-jih

Chi-tung
(Kyirong Dzong)

Jung-ka

Mustang

Baglung

Pyuthan

Nuwakot

Pokhara

Gurkha

Salyan

Jumla

Simikot

Baitadi

Dandeldhura

Silgarhi-Doti

Dhangarhi

Kaurlala
Ghat

Nepalganj

Jarwa

Barhni

Bahraich

Gonda

Faizabad

Lucknow

Kanpur

Shahjahanpur

Bhairawa

Butanwa

Bhikhna Thori

Raxaul

Birganj

Amlekhganj

Bhimphedi

Patan

KATHMANDU

Bhadgaon

Kodari

Rasua
Garhi

Nawakot

Ramechhap

Okhaldhunga

Janakpur

Jaynagar

Dhapkuta

Rajbiraj

Biratnagar

Ilam

Darjeeling

Gorakhpur

1

Land and Agriculture in Nineteenth-Century Nepal

This study seeks to discuss the structure of the agrarian society of the Kingdom of Nepal during the nineteenth century, the peasant's role in that society, and the conditions under which he was able to obtain land for cultivation. Economic development, in the modern sense, requires the investment of capital, which must be accumulated mainly through savings. In countries such as Nepal where the agricultural sector has always been large, relative to the total economy, agriculture must be a major source of such savings. The study will, therefore, examine to what extent the agrarian structure of nineteenth-century Nepal was conducive to the growth of savings and to their productive investment. In particular, it will make an attempt to identify the institutional mechanism through which the economic surplus generated by the Nepali peasant was extracted from him.

During the nineteenth century, Nepal was largely a peasant society. Agriculture was the predominant economic activity, both in terms of total national product and the working population. The typical unit of agricultural production was the peasant household, producing both for its own consumption and for sustaining the state, landowners, and other privileged groups in the society. The predominance of agriculture based on the peasant household imparted a basic unity to the economy of nineteenth-century Nepal, which makes possible a number of generalizations about correlation between the structure of agrarian institutions and the level of economic development. This unity was accentuated by the basic uniformity of tenurial forms, particularly after political unification was achieved during the latter part of the eighteenth century. This did not, however, mean that the stru cture of the agrarian society was uniform throughout the country. On the contrary, it presented a motley pattern in almost all important respects. The main reasons for this diversity were geographical and historical.

The Geographic Background[1]

Nepal is a land of unmatched diversity of climate and topography. It encompasses almost all the climatic zones of the world, and ranges in altitude from the world's highest point at the peak of Mt. Everest (Sagarmatha) in the north to only a hundred feet above sea level in the south. It is not surprising, therefore, that agrarian conditions should present a diversity out of all proportion to the total surface area of 141,000 sq. km.

One-sixth of this area is situated in the Tarai, a narrow tract of level, alluvial terrain comprising the Ganges plain. Situated between the Indian frontier in the south and the Siwalik hills in the north, the Tarai is about 100 m. above sea level and is only about 45 km. in width. Between the Siwalik hills and the Mahabharat range in the north is situated the inner Tarai, with a topography similar to that of the Tarai, but with somewhat gravelly soils. This region comprises approximately one-tenth of the surface area of the Kingdom.

Between the Mahabharat range and the main Himalaya mountains are situated the midlands, a complex of hills and valleys 60 to 100 km. in breadth and covering much of the length of the country, at elevations ranging from 600 to 2,000 m. above sea level. Kathmandu Valley, comprising an area of about 600 sq. km., is located in the center of this region. It accommodates the capital city of Kathmandu. The main Himalayan range, situated some 80 km. north of the Mahabharat range, is largely an arctic waste. It contains at least 250 peaks of more than 6,000 m. in altitude along a distance of about 800 km. No vegetation is possible in most of the Himalayan region, the landscape is wild and desolate, and no human habitation is possible in the upper reaches. In western and central Nepal, some areas of the Kingdom are situated north of the main Himalayan range.

The Historical Background

These diversities of climate and topography are compounded by differences in the historical background. A meaningful interpretation of Nepal's agrarian history is, therefore, not possible without an understanding of the political divisions that existed during the period before its unification in the latter part of the eighteenth century. The territories now comprising the Kingdom of Nepal were then under the control of at least 60 petty principalities. Kathmandu Valley had been divided into three principalities—Kathmandu, Bhadgaun, and Lalitpur—at least since the middle of the fifteenth century. To the east were situated the Kingdoms of Chaudandi and Vijayapur, which controlled the hill regions south of Tibet, and, in addition, the modern Tarai districts of Jhapa, Morang, Saptari, Siraha, Mahottari, Dhanusha and Sarlahi. The Kingdom of Makwanpur, situated south of Kathmandu Valley, comprised the Tarai districts of Bara, Parsa, and Rautahat, parts of Chitaun in the inner Tarai, and some territories in the adjoining hills.

Political divisions in the regions situated west of Kathmandu Valley are more important from the viewpoint of land and agriculture during the nineteenth century. These regions were then under the control of a number of principalities belonging to two major groups: Chaubisi and Baisi. Chaubisi was the term used to denote a group of 24 principalities situated west of the Trishuli river. This region roughly encompasses 17 of the modern administrative districts of Nepal: Dhading, Nuwakot, Tanahu, Lamjung, Syangja, Kaski, Parbat, Gulmi, Argha-Khanchi, Palpa, Myagdi, Baglung, Rukum, Rolpa, Salyan and Pyuthan, as well as the district of Gorkha, the last independent principality founded in the western hill region, which was never considered a constituent of the Chaubisi group. During the latter part of the eighteenth century, Gorkha laid the foundation of the modern Kingdom of Nepal through territorial expansion in the east, west and south. Agrarian systems and institutions prevalent in Gorkha and the Chaubisi region were then gradually extended to Kathmandu Valley and the adjoining areas of the eastern hill region. In the present study, we shall use the term central hill region to denote all these areas. This region was situated between the Bheri river in the west and the Dudhkoshi river in the east.

Farther west, in the region situated roughly between the Bheri and Mahakali rivers, there were 22 principalities, collectively known as Baisi, which had once formed constituent units of the Kingdom of Jumla. Fifteen of Nepal's modern districts comprised the Baisi region: Dailekh, Jajarkot, Kalikot, Jumla, Bajhang, Bajura, Doti, Achham, Darchula, Baitadi, Dandeldhura, Mustang, Dolpa, Mugu and Humla. During the early nineteenth century, the Baisi region comprised only three administrative divisions: Jumla, Doti and Dullu-Dailekh, and three feudatory principalities: Bajhang, Bajura, and Jajarkot.

The political history of the modern Kingdom of Nepal starts during the 1760s, when Gorkha, under the leadership of King Prithvi Narayan Shah, conquered the three Kingdoms of Kathmandu Valley and the Kingdoms of Makwanpur, Chaudandi, and Vijayapur. The capital of the new Kingdom was shifted from Gorkha to Kathmandu in 1768. During the subsequent two decades, Kathmandu extended its military conquestst oward the east and the west, thereby ending the existence of the petty principalities mentioned above, either through outright conquest or through conversion to feudatory status. The process of military expansion culminated in the acquisition of territories along a distance of approximately 1,300 miles from the Tista river in the east to the Sutlej river in the west by the first decade of the nineteenth century. A large part of this territory was, however, surrendered to the East India Company as a result of war during 1814-16. Some of these territories were subsequently restored to Nepal, first in December, 1816 and then in November, 1860. The Kingdom acquired its present frontiers in the south as a result of these territorial adjustments, while the northern frontier had been more or less stabilized after the Nepal-China war of 1791-92.

Agricultural Regions

The political division of the hill regions west of Kathmandu Valley into the Chaubisi and Baisi regions before their incorporation into the Kingdom of Gorkha may have had its basis on geographical and economic factors. It is certainly no fortuitous coincidence that the Chaubisi region, together with Kathmandu Valley, was conterminous with what is now described as the central hill-farming region, the second most important agricultural region of Nepal. This region comprises broad, well-watered mountain valleys with deep, rich soils and carefully-terraced hillsides. Most areas in this region are situated at altitudes of 600 to 2,000 m. and are ideally suited for the cultivation of rice. In contrast, the Baisi region, comprising the far-western hill-farming region extending from Pyuthan to Dandeldhura, presents, on the whole, a composite landscape of an elevated plateau at altitudes generally exceeding 2,000 m. Because of exessively steep slopes, unsuitable soils, bad drainage, high elevation, low rainfall and other factors, large areas in the far-western hills either cannot be cultivated or support only marginally productive cultivation. The major crops of this region are maize, rice, millet and wheat, in that order. Rice and wheat, in particular, can be cultivated only in the southern sections of the river valleys, comprising less than one-quarter of the total cultivated area. The Tarai region is, of course, of greater importance in the economy of Nepal. Francis Hamilton, a member of the first British mission to Nepal in 1802-3, has recorded that during the early nineteenth century, the Tarai region had extensive tracts of forests containing valuable timber, as well as "much poor high land overgrown with trees and bushes of little value" and "a very large proportion of rich land."[2] Hamilton also noted that the rulers of the principalities which controlled the Tarai region before it was conquered by Gorkha "were so much afraid of their neighbours, that they did not promote the cultivation of this low land." The Gorkhalis, however, "being more confident, have cleared much of the country, although still a great deal remains to be done." Large quantities of grain were exported to India from the Tarai region, and, he goes on to say, "were property somewhat more secure, this territory is capable of yielding considerable resources."[3] The eastern Tarai, in particular, comprising the districts of Parsa, Bara, Rautahat, Mahottari, Saptari and Morang, has been described as the outstanding agricultural region of Nepal. It is a hot and humid area, with an annual rainfall of more than 80 in. and rich alluvial soils, which permit the cultivation of two or three crops per year. Its high revenue-yielding potential was one reason for its incorporation into the Kingdom of Gorkha during the mid-1770s.

The far-western Tarai region, comprising the "Naya Muluk" territories of Banke, Bardiya, Kailali and Kanchanpur which Nepal had lost to the East India Company in 1816, but regained in 1860, is a poorer agricultural area. Rainfall averages 40 in. per year and agricultural yields are consequently much lower than in the eastern Tarai. During the nineteenth century, this region had vast tracts of forest lands, whereas cultivated tracts were few and dispersed. Although the inner Tarai region also possessed considerable development potential, it appears to have remained practically a desert during the nineteenth century. The malarial climate was one constraint on its development. According to Hamilton: "The chief reason of the desert state of this part of the country, seems to be its extreme unhealthiness, and this again, in a great measure, in all probability, depends on the want of cultivation."[4] During the eighteenth and nineteenth centuries, the valleys of the inner Tarai region were left undeveloped also for considerations of national security.[5] As Oldfield, British Residency surgeon during the 1850s, noted:

In Nipal the *dhuns* (i.e., the valleys of the inner Tarai region) have been mostly allowed to fall into a state of jungle, and are consequently clothed with forests in sal and cotton trees, and are inhabited only by wild beasts. The Nipalese are averse to the 'clearing' of these forests, as they look upon the malarious jungle at the foot of their hills as the safest and surest barrier against the advance of any army of invasion from the plains of Hindustan.[6]

It was in accordance with the policy of developing the inner Tarai region as a barrier against external aggression that the government of Nepal, after its defeat in the 1814-16 war with the East India Company government in India, adopted the policy of discouraging settlement in the central inner Tarai.[7] The policy continued to be followed with the objective of isolating Kathmandu Valley even when the development of friendly relations with the British had alleviated fears of aggression from the south.[8]

The diversity of climate and topography, compounded by the accidents of history, have inevitably fostered a diversity of agrarian systems and institutions, as well as of economic activity, and hence disparities in the stages of economic development in different parts of the country. These disparities have never been adequately studied, but it may be sufficient in the present context to record that while agriculture was important almost everywhere, in large areas of the Tarai manufacturing and commercial activity appears to have been conducted on a scale that reminds one of the European accounts of India during the early sixteenth century. Indeed, the Tarai region, thanks to its proximity to northern India, contained the most developed market economy in the country. The hill region, particularly in the west, produced large quantities of copper, iron and other metals of high quality. At the same time, many areas in the hill region comprised a primitive, tribal social and economic order, while in the northern Himalayan zones, nomadic groups combined trade with animal husbandry and seasonal cultivation. On the other hand, the principalities of Kathmandu Valley, which was situated on the traditional trade route between India and Tibet, presented a cross between the warring city-states of ancient Greece and the trading towns of southern Europe in the later middle ages.

A Cereal-Farming Economy

As mentioned previously, the objective of the present study is to identify the institutional mechanism through which the economic surplus generated by the Nepali peasant during the nineteenth century was extracted from him. The nature of that mechanism was different for different categories of agricultural production. The Nepali peasant during the nineteenth century produced not only such staple food crops as rice, wheat, maize and millet, but also a number of commercial crops, including cotton and cardamom in the hill region, and indigo, opium, sugarcane and tobacco in the Tarai. The government followed different fiscal policies for food and commercial crops. Taxation constituted the chief method whereby economic surpluses were extracted from peasants who produced food crops. In the case of commercial crops, on the other hand, that objective was attained through a system of state trading and monopolies. The present study will confine itself to the taxation policies followed towards the food-producing peasant. Policies followed towards peasants who produced commercial crops will hopefully form the subject-matter of another volume. This should not, of course, give rise to the impression that Nepali agrarian society during the nineteenth century consisted of two separate classes of peasants, one producing food crops, and the other commercial crops.

Rice-Lands and Homesteads

It is necessary at this stage to describe the categories of agricultural lands that were used for the production of the staple food crops—rice, wheat, maize and millet—in the central hill region and the Tarai. These categories are rice-lands and unirrigated lands, including highlands and hillside lands, which usually formed a part of the peasant's homesite. Rice was the main crop in both the central hill region and the Tarai. Hamilton has estimated that "on the whole one-half of the cultivation among the mountains may be said to consist in transplanted rice."[9] The importance of rice in the agricultural economy is apparent from the fact that agricultural lands were classified on the basis of their suitability for rice production. The term *khet* in the central hill region, and *dhanahar* in the Tarai, denoted lands which were suitable for the cultivation of rice. *Pakho* or *bhith*, on the other hand, denoted unirrigated lands which were suitable for use as homesites, and for the cultivation of maize, millet and other dry crops which do not require flooding of fields.[10]

Rice grows only on lands which can retain rainwater, or can be irrigated through channels cut from streams or springs. Rice lands are, therefore, situated on the banks of streams and rivers, as well as on hill-sides which can be terraced and irrigated through artificial means. Such lands are, ordinarily, more productive than the unirrigated lands contained in homesteads. The regular availability of irrigation facilities permits the cultivation of more than one crop from the same field each year, and, moreover, minimizes risks of crop failure because of drought. Rice lands are, therefore, of considerable importance in the economy of the peasant household.

Homestead lands, on the other hand, are usually situated on hillsides. The unirrigated lands contained in homesteads yielded coarse grains such as millet and buckwheat, which probably constituted the staple diet of the peasantry. As Kirkpatrick, the first Englishman to visit Nepal in 1793, has recorded, "these articles are chiefly consumed by the husbandmen themselves and others among the lower classes of people."[11] The introduction of maize during the early eighteenth century greatly increased the importance of unirrigated lands in the rural economy.[12] Such lands then began to yield at least two crops a year: maize during April-July and millet during August-December. The volume of food production increased along with improved peasant productivity. If the experience of other parts of the world is any guide, it is fairly certain that the population of Nepal witnessed a significant increase during and after the mid-eighteenth century as a result of this development.[13]

There existed differences also in the systems followed for the assessment of taxes on rice lands and homesteads. As a rule, taxes on rice lands were assessed on each unit of area, usually the *muri* in the hills, and the *ropani* in Kathmandu Valley, whereas on home-steads taxes were assessed on the roof, as well as on lands com-prising the homestead, on the basis of a rough estimation of the number of ox-teams required to cultivate it. This difference in the tax-assessment systems followed for rice lands and homesteads had a direct impact on revenue. Because the rice-land tax was based on the area, the same area of land yielded the same revenue from year to year, irrespective of the size of the village population. In con-tradistinction, homestead tax revenue increased even if the area occupied by homesteads remained unchanged when existing home-steads were split as a consequence of partition or subdivision. Also, homesteads were often depopulated because of death, emigra-tion, or other reasons. The amount of homestead-tax revenue, con-sequently, changed from year to year.

The distinction between rice lands and other categories of agri-cultural lands was meaningful from the viewpoint of land tenure and taxation to a greater extent in the hill region than in the Tarai. This is so mainly because of topographical conditions and the settle-ment patterns. In the hill region, homesteads are usually of the dis-persed type. A part of the homestead is used for residential pur-poses; the remaining area is used to grow dry crops. A homestead is thus not only a residential unit but also a unit of agricultural pro-duction and, consequently, of tax assessment. In the Tarai region, where topographical diversities are not so marked as in the hill regions, homestead lands were not regarded as a separate category for purposes of taxation. The settlements are of the compact type[14] and peasants' dwellings usually occupy sites which contain little space beyond what may be needed for accommodating cattle or for use as kitchen gardens. Consequently, "the tenantry pay no ground rent for their houses."[15]

The Cropping Pattern

From the viewpoint of climate and topography, large areas of agricultural lands in both the central hill region and the Tarai are capable of yielding more than one crop a year. During the nine-teenth century, however, multi-cropping appears to have been sel-dom practised, mainly because of the depredations of stray cattle and wild animals and the scarcity of irrigation facilities.

Agricultural holdings usually consisted of small and dispersed fragments. The erection and maintenance of fences required capital investment on a scale which peasants could seldom afford. Consequently, crops remained exposed to the depredations of cattle. The observations made by Hamilton on the Indian territories adjoining Morang district in the eastern Tarai region during 1809-10 may be regarded as more or less applicable to the Nepali side of the border as well.[16]

> In most places there is no sort of attempt to close anything but the yard which surrounds the hut . . . . In many parts kitchen gardens are quite defenceless, or are guarded merely by a few dry bushes . . . . The want of fences is a great evil, and the cattle commit uncommon depredations . . . . The people who tend the cattle seem to be sent rather with a view to prevent them from straying, then to keep them from destroying the crops, at least I saw many instances of a most culpable neglect. I have here very seldom observed cattle tethered, which in an open country is a very useful practice.

Nor were stray domestic cattle the only menace to crops. The depredations of wild animals, particularly elephants, was another reason why multi-cropping was rarely practised in the Tarai region. Again according to Hamilton:[17]

> In the dry season the elephants return to the lower ranges of hills; but in the rainy season they abandon these forests, and are then very destructive to the crops, which, indeed, prevents natives from being so attentive to the cultivation of rice as they otherwise would be, so that, although the country is best adapted for the culture of this grain, the farmers content themselves chiefly with winter crops of wheat, barley, and mustard.

In the hill region, cattle were let loose in the fields during winter to graze on the stubble after the rice-crop was harvested, thereby hindering the cultivation of wheat, barley and other winter crops. Hamilton noted that although rice lands in that region could also grow a winter crop, "in most places this is most judiciously omitted."[18] The practice may have been described as judicious because the droppings of cattle operated as fertilizer, thereby augmenting the yield of the rice crop during the monsoon season.

Irrigation

There were, of course, other reasons also for the peasant's reluctance to cultivate winter crops. Oldfield has recorded that only one crop in the year was grown in Nuwakot, because of "the excess of moisture in the *byasis* (i.e., river valleys) and the total want of artificial irrigation in the *tars*" (i.e., highlands)."[19] Indeed, there is little evidence to show that irrigation facilities were much developed in any part of the country. According to another contemporary British account:[20]

> Although the country is everywhere interspersed with streams, which might be made available for the purpose of irrigation, and the people are perfectly well aware of their great value in this respect, yet from not possessing the means of conducting the water to the high grounds, a large portion is entirely wasted and land which might otherwise afford a profitable return to the cultivator remains unutilized.

It may be correct, therefore, to presume that irrigation facilities remained undeveloped because of the peasant's inability or unwillingness to finance irrigation schemes. Construction of new irrigation channels was usually a costly enterprise. Moreover, such channels often passed through the fields of several peasants, and this gave rise to disputes and litigation.[21] Their construction required cooperative effort which only rich and influential farmers were capable of organizing.

A system of state-operated irrigation canals existed in several districts of the western hill region, including Kaski, Nuwakot, Palpa,[22] and Syangja,[23] during the nineteenth century. These canals were maintained and repaired through the compulsory labour of the peasants whose rice fields they irrigated.[24] The government appointed superintendents to supervise such maintenance and repair operations and arrange for the equitable and maximum utilization of the available irrigation facilities.[25] No information is available regarding the area of rice lands irrigated by these state irrigation canals. It appears likely, however, that the area was not large. In the Pokhara area, for instance, where there were at least four such irrigation canals, Oldfield noted that "a large part of the valley is under little or no cultivation."[26]

Agriculture in Kathmandu Valley

A state-operated irrigation system had long existed in several areas of Kathmandu Valley also.[27] Moreover, there existed customary arrangements for the protection of crops from stray cattle. Functionaries were appointed in each village to prevent such cattle from destroying crops. These functionaries were held personally liable if crops were destroyed by stray cattle, or if paddy was stolen from the threshing ground.[28] Thanks to these facilities, as well as a favourable climate and rich soils, agriculture appears to have been much better developed in Kathmandu Valley than in most other parts of the country. Contemporary European observers have given accounts of the developed state of agriculture in this region. Sir Richard Temple, who visited Nepal in 1876, wrote:[29]

> The cultivation of the Nepal Valley is blessed with unequalled advantages, and is carried on with utmost industry. In May we found a waving harvest of wheat awaiting the sickle, and I was told that almost all these lands had already yielded an equally good rice harvest within the agricultural year, and that many of the fields would yet yield special crops, pepper, vegetables, and the like. In short, most of the lands yield two harvests in the year, and some yield even three.

Temple admitted that the chemical quality of the soil "must be excellent" to permit such multi-cropping, but noted, at the same time, that "one special cause of the fertility is the artificial irrigation from the countless streams and streamlets from the neighbouring hills."[30] Dr Daniel Wright, British Residency surgeon during the early 1860s, similarly noted that in Kathmandu Valley "every available scrap of ground is cultivated, the hill-sides being terraced wherever water can be obtained for irrigation."[31] He also noted that "most lands yield two crops a year, and from some even three crops are obtained," and that "there is no grazing ground except at the foot of the hills."[32]

Kathmandu Valley, however, represented an island of high agricultural productivity against a general background of inefficient and extensive farming practices. It is true that little information is available about the techniques followed in the cultivation of cereal crops in the central hill region and the Tarai. Kirkpatrick[33] and Hamilton[34] have given brief descriptions of such techniques in

Kathmandu Valley, but it would be unwarranted to make any generalizations on the basis of techniques followed in this area. At the same time, their accounts leave us in no doubt that agricultural techniques have long remained unchanged even in Kathmandu Valley. The experience of other parts of the country could, therefore, scarcely have been better. Indeed, there is no evidence that any technological innovation or improvement was adopted in agriculture in any part of the country at any time during the nineteenth century.

NOTES

[1]The sections on geographical divisions, agricultural regions, and crops are based on the following sources: Pradyumna P. Karan et al., *Nepal: A Physical and Cultural Geography*, Lexington, University of Kentucky Press, 1960; Toni Hagen, *Nepal; The Kingdom in the Himalayas*, Berne, Kimmerley and Frey, 1961; and Charles McDougal, *Village and Household Economy in Far-Western Nepal*, Kirtipur, Tribhuwan University. n. d. (1968).

[2]Francis Hamilton, *An Account of the Kingdom of Nepal*, (reprint of 1819 ed.) New Delhi, Manjusri Publishing House, 1971, p. 62.

[3]*Ibid.*, p. 64.

[4]*Ibid.*, p. 69.

[5]*Ibid.*, p. 198.

[6]H. A. Oldfield, *Sketches from Nipal* (reprint of 1880 ed.) Delhi, Cosmo Publications, 1974, p. 47.

[7]"Order regarding Evacuation of Cultivated Lands and Settlements in the Kamala-Chitaun Region," Aswin Sudi 11, 1874 (September 1817), *Regmi Research Collection*, Vol. 43, p. 38.

The *Regmi Research Collection* (hereafter *RRC*) is the private collection of the author, comprising copies of unpublished documents obtained mostly from the Lagat Phant (Records Office) of the Department of Land Revenue in the Finance Ministry of His Majesty's Government.

[8]"Order regarding Evacuation of Cultivated Lands and Settlements in the Bhimphedi-Hitaura Region," Aswin Badi 13, 1931 (September 1874), *RRC*, Vol. 68, pp. 36-38.

[9]Hamilton, *op. cit.*, p. 75.

[10]Oldfield, *op. cit.*, p. 91; Richard Temple, "Remarks on a Tour Through Nepal in May, 1876" in *Journals Kept in Hydrabad, Kashmir, Sikkim, and Nepal*, London, W.H. Allen, 1887, Vol. 2, p. 227.

[11]William Kirkpatrick, *An Account of the Kingdom of Nepaul*, (reprint of 1811 ed.) New Delhi, Manjusri Publishing House, 1969, p. 94.

[12]Mahesh C. Regmi, *A Study in Nepali Economic History, 1768-1846*, New

14 Thatched Huts and Stucco Palaces

Delhi, Manjusri Publishing House, 1971, p. 17.

[13]Jordi Nadal. "The Failure of the Industrial Revolution in Spain, 1830-1914," in Carlo M. Cipolla, *The Fontana Economic History of Europe*, Part II: *The Emergence of Industrial Societies*, Glasgow, William Collins Sons & Co. Ltd, 1975, p. 533. According to Nadal: "In the Danube basin the rural population doubled in a very short time after the introduction of maize."

[14]Karan, *op. cit.*, p. 56; L.R. Singh, *The Tarai Region of U.P.*, Allahabad, Ram Narain Lal Beni Prasad, 1965, p. 50.

[15]Hamilton, *op. cit.*, p. 154.

[16]Francis Buchanan (Hamilton), *An Account of the District of Purnea in 1809-10*, Patna, Bihar and Orissa Research Society, 1928, pp. 424-25.

[17]Hamilton, *An Account af the Kingdom of Nepal, op. cit.*, p. 63.

[18]*Ibid.*, p. 73.

[19]Oldfield, *op. cit.*, p. 154.

[20]Orfeur Cavanagh, *Rough Notes on the State of Nepal*, Calcutta, W. Palmer, 1851, p. 102.

[21]Cf. "Order to Gorkha Adalat Regarding Complaint against Bhanubhakta Pandit and Others," Kartik Badi 14, 1921 (October 1864), *RRC*, Vol. 49, pp. 402-4.

[22]"Appointment of Rupanarayan as Caretaker of Irrigation Channels in Kaski District," Poush Sudi 4, 1902 (December 1845), *ibid.*, Vol. 26, p. 49; "Appointment of Maniram Koirala as Caretaker of Irrigation Channels in Kaski, Nuwakot, and Palpa," Marga Sudi 3, 1903 (November 1846), *ibid.*, Vol. 26, pp. 166-67.

[23]"Appointment of Bahadur Khatri and Karna Singh Khatri as Caretakers of Irrigation Channels in Kaski and Syangja," Marga Sudi 3, 1903 (November 1846), *ibid.*, Vol. 26, pp. 163-65.

[24]"Order to Tenants Cultivating Irrigation Lands in Lamachaur, Kaski District," Magh Badi 14, 1892 (January 1836), *ibid.*, Vol. 27, p. 360.

[25]"Appointment of Rupanarayan as Caretaker of Irrigation Channels in Kaski District," 1845 (see n. 22 above).

[26]Oldfield, *op. cit.*, p. 46.

[27]Cf. Chittaranjan Nepali, *Janaral Bhimsen Thapa ra Tatkalin Nepal* (General Bhimsen Thapa and Contemporary Nepal), Kathmandu, Nepal Sanskritik Sangh, 2013 (1956), p. 209; "Royal Order to Dhalwas of Panga and Kirtipur," Kartik Badi 5, 1851 (October 1794), *RRC*, Vol. 5, pp. 292-93.

[28]"Order regarding Collection of Sahanapal and Other Levies in Kathmandu and Kirtipur," Ashadh Sudi 6, 1850 (June 1793), *ibid.*, Vol. 5, pp. 250-51; "Regulations regarding Protection of Crops in Kathmandu," Bhadra Badi 14, 1900 August 1843), *ibid.*, Vol. 33, pp. 504-8.

[29]Temple, *op. cit.*, p. 251.

[30]*Loc. cit.*

[31]Daniel Wright (ed.), *History of Nepal*, (reprint of 1877 ed.) Kathmandu, Antiquated Book Publishers, 1972, p. 46.

[32]*Loc. cit.*

[33]Kirkpatrick, *op. cit.*, p. 100.

[34]Hamilton, *An Account of the Kingdom of Nepal, op. cit.*, pp. 221 ff.

2

Politics and Government

The political system of any society is a basic part of its organization[1] and thus has a profound impact on its economic life. The
question of whether economic conditions have shaped the political
system of the society or *vice versa* can hardly have a definitive
answer. Nevertheless, there can be no doubt that an adequate understanding of the political system is essential for a meaningful
study of economic history. Our study of agrarian relations in Nepal
during the nineteenth century, therefore, must be preceded by a
brief discussion of the historical background of contemporary
politics and government.

Political Developments during the Nineteenth Century

A brief description of Gorkha's campaign of territorial expansion during the latter part of the nineteenth century, which culminated in the political unification of the territories now comprising
the Kingdom of Nepal, has already been given in the preceding
chapter. Notwithstanding political unification, the new Kingdom
was unable to enjoy political stability, mainly because of interne-
cine conflict among members of the nobility and even those of the
royal family. Matters came to a head in early 1799, when the King,
Rana Bahadur Shah, a grandson of Prithvi Narayan Shah, abdicat-
ed in favour of an infant son, Girban Yuddha Bikram Shah, and
went into voluntary exile in India. He returned to Nepal five years
later, and assumed charge of the administration, but was assassinated in April 1806. Bhimsen Thapa, a member of the nobility

who had remained loyal to Rana Bahadur Shah, then became Prime Minister. For 31 years, from 1806 to 1837, he ruled Nepal with virtually unchallenged authority. He was able to retain his position even after the King, Rajendra Bikram, a son of Girban, had attained majority.

In 1837, Bhimsen Thapa was dismissed and imprisoned, and eventually committed suicide in jail. For nearly nine years thereafter, Nepal was a victim of political instability at the hands of factions headed by King Rajendra Bikram, his two queens, and the Crown Prince, Surendra Bikram Shah, each with supporters among the nobility. In May 1845, the new Prime Minister, Mathbar Singh Thapa, was assassinated after nearly two years in office. A four-member government was then formed. One of the members of that government was Jung Bahadur Kunwar. Political conflict among the nobility continued, however, culminating in a massacre of leading members of the important political families on 14 September 1846, and the flight or banishment of several others.

The Rise of the Rana Family

On 15 September 1846, Jung Bahadur was appointed Prime Minister of Nepal. He laid the foundation of a political system which survived until 1951 notwithstanding occasional inter-familial conflicts and political conspiracies. The chief features of that system was the political neutralization of the King. In fact, all the three Shah Kings who reigned during the Rana period, Surendra (1847-81), Prithvi (1881-1911) and Tribhuwan (1911-55), were confined by the Ranas to the royal palace and kept under strict surveillance.

Jung Bahadur had seven brothers who had all rendered great assistance in his rise to power. It was, therefore, necessary for him to devise a system under which their personal political ambitions could be satisfied. A royal charter promulgated in 1856 accordingly provided that each of Jung Bahadur's brothers should in turn become Prime Minister in the event of a vacancy in that office.[2] Succession thereafter went in order of seniority to Jung Bahadur's sons and to the sons of his brothers. Bam Bahadur, next in seniority among the brothers of Jung Bahadur, thus became Prime Minister when Jung Bahadur voluntarily relinquished office in July 1856. He died in May 1857, and Jung Bahadur reassumed the post of Prime Minister,

Jung Bahadur was succeeded by another of his brothers, Ranoddip Singh, when he died in 1877. The Rana family subsequently became divided into two hostile factions: the Jung faction, consisting of the 10 sons of Jung Bahadur, and the Shumshere faction, consisting of the 17 sons of his youngest brother, Dhir Shumshere. In November 1885, the Shumshere faction, led by Bir Shumshere, assassinated Prime Minister Ranoddip Singh. Bir Shumshere then declared himself Prime Minister. He instituted a new roll limiting the right of succession to his own brothers and sons in order of seniority, thereby disfranchizing the sons of Jung Bahadur and his brothers. Bir Shumshere's 15-year rule (1885-1901) almost coincided with the end of the nineteenth century.

The External Scene

These internal political changes occurred about the same time as far-reaching changes in the external political situation. Nepal's defeat in the 1814-16 Nepal-British War had created a crisis of national identity and objectives. Efforts to enlist assistance from China to avenge this defeat had proved consistently unsuccessful. Indeed, China itself had been badly humiliated by the Opium Wars and weakened by internal rebellions, and so was hardly in a position to help Nepal. Kathmandu realized that China was neither able nor willing to help it in any future war against the British. The extent of China's impotence became clear during the 1855-56 Nepal-Tibet War, which it was able neither to prevent nor to influence in Tibet's favour.

The shift in the balance of power was by no means the sole factor that brought Nepal and the British closer to each other. Of perhaps greater importance was the growing contact between the two sides after the middle of the nineteenth century. Prime Minister Jung Bahadur paid a visit to England in 1851, and personally led an army to India to help the East India Company crush the 1857 rebellion. Such contacts and cooperation inevitably led to mutual understanding and trust and thereby to a basic change in Nepal's foreign policy. Nepal now veered away from China and tilted towards the British.

The British success in suppressing the 1857 rebellion made it an unchallenged power in the Indian sub-continent. It also changed the entire basis of British rule. After power was taken over by the British Crown from the East India Company, India was no longer ruled by a gang of adventurers, frantic to enrich themselves. In the middle of the eighteenth century the British were still organized for commerce and plunder in the East India Company and controlled no more than a small fraction of Indian territory. By the middle of the nineteenth century, however, they had become in effect the rulers of India, organized in a bureaucracy proud of its tradition of justice and fair dealing.[3]

Benefits of Nepal-British Friendship

Cordial relations with the British brought several important benefits to Nepal. The accretion of territory in the far-western Tarai was the most important of these benefits. Under the 1816 treaty, Nepal had surrendered to the East India Company the whole of the Tarai areas situated between the Kali and Rapti rivers in the west. These territories were restored to Nepal in November 1860 "in recognition of the eminent services rendered to the British government by the State of Nepal" during the 1857 rebellion.[4] Nepal thereby acquired approximately 2,850 sq. miles of territory in the present far-western Tarai districts of Banke, Bardiya, Kailali and Kanchanpur. Jung Bahadur's policy of friendship towards the British thus helped to recoup a small part of the territorial losses that Nepal had sustained as a result of the 1814-16 war. The newly-acquired territories contained valuable forests and extensive tracts of cultivable lands.

Moreover, generally speaking, the internal political boundaries of India became fixed after 1857. The native princes were thereafter no longer afraid of expropriation, and so identified their interests with those of the British. Neither Nepal nor the British now had aggressive designs on the territories of each other, with the result that there was no basic conflict in their interests, and hence no rationale in the policy of "peace without cordiality"[5] that had characterized the period after the 1814-16 war. Regulations promulgated by the government of Nepal for different districts of the Tarai region before and after 1857 clearly reflect the changed situation. Throughout the late eighteenth-early nineteenth centuries, local authorities in that region were told:

If Chinese and English troops violate the borders and kill or loot our people, take appropriate steps to defend our territories. Refer the matter to us for instructions if there is time, and act according to such instructions. If not, take appropriate steps to defend our territories and repulse the enemy.[6]

Ten years later, however, the same authorities were ordered to take steps to repel external aggression only with Kathmandu's approval,[7] thereby implying that the government did not anticipate any threat to national security that might necessitate urgent action through local initiative. The government of Nepal was thereafter able to pursue its policies of reorganizing the district administration as well as of speeding up land reclamation and settlement in the Tarai region without any fear of external aggression.

The task of maintaining law and order in the Tarai region was facilitated also by the signing of an extradition treaty between Nepal and the British government in early 1855. The treaty required each government to extradite criminals guilty of "murder, attempt to murder, rape, maiming, thuggee, dacoity, highway robbery. poisoning, burglary and arson" who escaped into its territories.[8] The signing of that treaty, and the generally cordial relationship between the two governments that Prime Minister Jung Bahadur succeeded in establishing, also facilitated joint action in checking crime in the border areas. At the same time, officials from British India were forbidden to intrude into Nepali territory in pursuit of criminals, and Nepali officials too were directed not to intrude into British Indian territory for such purposes.[9]

The New Political Elite

The rise of the Rana family was the consequence of internal changes within the framework of the traditional political class system, not a case of vertical mobility. Throughout Nepal's post-1768 history, participation in the political process had become the exclusive prerogative of the Brahman and Chhetri families who had followed King Prithvi Narayan Shah from Gorkha to Kathmandu. Jung Bahadur belonged to one of the less influential sections of these families, which had distinguished itself at the middle echelons of the administration and the army rather than in the matrices of central politics. Jung Bahadur's emergence in Nepal's political scene cannot, therefore, be regarded as a departure from the political traditions of the kingdom.

This did not mean, however, that the Rana political system functioned along traditional lines. Inasmuch as it reduced the focal point of the traditional political system, the monarchy, to a nonentity, and eliminated other elements in the traditional nobility from the struggle for political leadership,[10] the Rana political system was as much against the traditions of the Shah dynasty as it was against the traditional political process.[11]

But even though the rise of the Ranas did not mark the entry of a new group into the traditional nobility, it certainly heralded the emergence of a new political elite, a small group within the political class "which comprises those individuals who actually exercise political power in a society at any given time."[12] Subsequently, royal orders were promulgated to give a special social status to the Rana family, and a legal status to its role as a political elite. These orders also recognized the Ranas' hereditary right to succession to the office of Prime Minister, and conferred a number of economic privileges on them.[13]

As mentioned above, Jung Bahadur belonged to one of the less prominent sections of the nobility that had followed King Prithvi Narayan Shah from Gorkha to Kathmandu. The Rana family originally bore the clan name of Kunwar, a Chhetri caste. Before Jung Bahadur became Prime Minister, it had no claim to a caste-status superior to that of the other sections of the traditional Gorkhali nobility. In May 1849, however, a royal order officially confirmed the Kunwars' claim to be the descendants of the Rana family of Chittor in India, and accordingly conferred on them the title of Rana. The Rana family thus attained a higher social status than the other sections of the nobility.

During the period from 1846 to 1856, Jung Bahadur functioned as Prime Minister in his individual capacity. The Rana family was, therefore, a mere *de facto* political elite which owed its status to the actual exercise of political power. Subsequently, it acquired that status through the exclusion by constitutional law of other political classes from political power, as well as through the formal institutionalization of its own privileges and obligations. The 1856 royal order formally limited succession to the Prime Ministership to members of the Rana family. Other sections of the nobility from among whom Prime Ministers had traditionally been appointed, such as Thapas, Pandes, and Chautariyas, were thereafter excluded from the ranks of the political elite. This order closed the doors of political power to the non-Rana political classes and relegated their role to oppositional politics aimed at the restoration of the pre-1846 power structure. The Rana family, comprising "the Vizier, and his brothers and sons" accordingly constituted the political elite that ruled Nepal until 1951.

In 1856, Jung Bahadur was designated as the Maharaja of Kaski and Lamjung, with special powers to impose or commute capital punishment, to appoint or dismiss government officials, to declare war or make peace with Tibet, China, and the British government or other foreign powers, to dispense justice and punishment to criminals, and to formulate new laws and repeal or modify old laws pertaining to the judicial and military departments of the government. The Maharaja was even authorized to prevent the King himself "from trying to coerce the nobility, the peasantry or the army, or from disturbing the friendly relations with the Queen of England and the Emperor of China." In the dual capacity of Maharaja of Kaski and Lamjung, and Prime Minister, the head of the Rana family exercised sovereign authority all over the Kingdom of Nepal.

Legislation was also enacted to confer a special status and specific obligations on the Rana Prime Minister and other members of the Rana family. For instance, they were prohibited "to accept tax-free land grants, except on forest lands, in the old territories of the Kingdom. However, they may accept tax-free grants in newly-acquired territories. They shall not accept any contracts for the collection of revenue, or be a partner in such contracts, or provide surety for persons who take up such contracts."[14] The law thus ensured a special status for the Rana family vis-a-vis other sections of the traditional nobility. It also prescribed that any attempt to assassinate the Rana Prime Minister or overthrow the rule of the Rana family should be regarded as an act of treason,[15] thereby giving the Rana family the status and dignity of a royal house.

Understandably, it is this parvenu nature of Rana rule that has motivated its condemnation in Nepali historiography, primarily by individuals or groups who had suffered directly or vicariously from it. The Nepali peasant in the nineteenth century obviously reacted to this event in a different way. Even before the rise of the Rana regime, social and political leadership was provided by Brahmans and Chettris, the descendants of early immigrants from northern India, and members of the local *khas* community who had succeeded in elevating their caste and social status. The mass of the peasantry played no role in politics. For them, it mattered little whether the individual or group who wielded power in Kathmandu belonged to the Thapa or to the Rana family. The question that needs to be answered in the context of the present study of landlord-peasant relations is, therefore, not the composition of the political elite, but the nature of the policies it followed in the administrative and economic fields.

Growth of a Civil Administration System

The political and administrative system that the Rana rulers inherited had one basic characteristic: sovereign authority and ownership rights in the land were vested in the king, but administrative functions, including the collection of revenue, were delegated to revenue farmers, land assignees and local functionaries. The central government, under that system, did not collect taxes and rents directly from the peasant. As a result, all that was left to the center was only "what (the local lords) choose, or think proper, to hand over to it."[16] Such a system naturally weakened the political and economic authority of the central government. Rana rule achieved a partial reversal of that trend. The Ranas succeeded in creating a civil administration which replaced the delegated authority of local administrators and revenue farmers. For the first time, distinct and separate organs of administration, devoted specially to fulfilling various administrative and government functions, emerged in Nepal.

The most necessary function of the newly-created civil administration was, of course, the collection of revenue. The Ranas laid the foundation of a system of revenue collection through salaried functionaries of the government, rather than by contractors or revenue farmers. The new district administrators were civil servants, not traders and financiers as was usually the case before. They were given military ranks and subjected to military discipline.[17] Most of them belonged to Kathmandu or the hill districts; hence their property could easily be impounded or confiscated, if necessary. Regulations were promulgated prohibiting them from acquiring lands or undertaking any trade in the areas where they were assigned.[18] Any official guilty of bribery or corruption was liable to be "dismissed from service, put in irons and brought to Kathmandu in a cage."[19] This was, indeed, a radical departure from the options available during the early years of the nineteenth century, when Kathmandu could do nothing but issue plaintive warnings to erring revenue farmers that "sin will accrue if unauthorized taxes are collected."[20]

The formation of a central office in Kathmandu to maintain a record of government employees of all ranks, as well as of their postings, transfers and promotions,[21] was another important step towards the evolution of a civil administration. This arrangement made it possible for the leave and other conditions of service of even district-level employees to be controlled directly from Kathmandu.[22]

The Rana government also appears to have paid due attention to the basic condition for the success of efforts to organize a civil administration: that "servants should be employed to keep a watch, or check, on other servants."[23] Accordingly, it established an independent judiciary for the first time in the history of Nepal, reorganized the audit system, and created permanent machinery to deal with corruption in the administration.

A central judiciary, known as the *Adalat Goswara*, was established in 1860. The judge of that court was granted full authority "to dispense justice according to the law, without fear or favour." He was expected to seek the sanction of the government only in matters regarding which the law contained no provisions.[24]

Although an office for the scrutiny of government accounts is said to have existed ever since the establishment of Gorkhali rule in Kathmandu,[25] it was reorganized in 1848 as a quasi-judicial body under General Badrinar Singh, a brother of Prime Minister Jung Bahadur, to audit accounts of government income and expenditure and dispose of cases of irregularities and corruption.[26] Detailed regulations were formulated for the maintenance of accounts of government revenue and expenditure, as well as for audit.[27]

The formation in 1870 of a high-level anti-corruption court, known as the *Dharma Kachahari*, was perhaps a measure of even greater significance. The judge of that court, who was appointed on a life-long tenure, was empowered to scrutinize complaints or evidence of bribery, injustice, etc., even against the Prime Minister and other senior members of the Rana family. He was required to comply with orders from even the King on any matter which was *sub judice* only after having those orders endorsed by the Prime Minister.[28]

The Legal Code

The infrastructure of jurisprudence which these administrative

arrangements necessitated was established through the enactment
of a legal code in early 1854 for the first time in the history of
Nepal. The code is, without doubt, one of the outstanding achieve-
ments of Rana rule. Its objective was "to ensure that uniform
punishment is awarded to all subjects and creatures, high or low,
according to (the nature of) their offense, and (the status of) their
caste."[29] For the most part, the code retained customary practices
relating to land tenure, as well as the traditional customs and
usages of different local or ethnic communities in the country. It
allowed an autonomous status to the customs and usages of each
community within the framework of the Rana legal and administrative system. In other words, the objective was "to regulate legal
activities in various spheres, thus regulating the entire systems of
social control these activities implied."[30] At the same time, the
code seems to have made an attempt to introduce reforms in a few
areas such as slavery, bondage, and the custom of *sati*.

From the viewpoint of the present study, two features of the
1854 legal code merit special attention: its constitutional character,
and its provisions for a civil administration system which could
exercise a certain degree of autonomy vis-a-vis the ruling elite.

The 1854 legal code contained several provisions which conferred definite powers and authority on both executive and judicial
officers in the regular exercise of their official functions. These
provisions debarred even the King or the Prime Minister from
encroaching upon the powers and authority specifically conferred
on executive and judicial officers.[31] The code prescribed that they
would not be held guilty if they disobeyed such orders, but that
obedience would be regarded as an act of guilt. Similarly:

> Government officers shall dispense justice according to the law.
> They shall not obey any order of the King, the Prime Minister,
> or the government to dispose of cases contrary to the provisions
> of the law. They shall not be punished on the ground that they
> have not complied with such orders.[32]

Indeed, the code laid the foundation of a constitutional system of
government in Nepal by prescribing that "everybody, from (the
King and other members of the royal family) to a ryot, and from
the Prime Minister to a clerk, shall comply with its provisions.[33]

Moreover, the 1854 legal code regulated administrative procedures and conferred certain rights on the citizen vis-a-vis the administration. For the first time in the history of Nepal, regular procedures were defined for different branches of the administration, thereby minimizing the scope for individual discretion. Government officials were required to specify the law and its particular section under which they made their decisions and judgments.[34] A definite procedure was laid down also for filing complaints against government officials and functionaries.[35] Anybody could now claim that the judgment pronounced on his case was at variance with the provisions of the code. The promulgation of the code also expedited administrative procedures, for no reference to the government was permitted in cases where power had been delegated.[36]

Impact on the Political System and the Administration

Nevertheless, neither the constitutional aspects of the 1854 legal code nor the autonomy that it sought to confer on the administration appears to have had a long-term impact on Nepal's political system and administration. Legislation alone could not circumscribe the reality of the Rana Prime Minister's absolute authority. There were no constitutional safeguards to ensure that he actually complied with the spirit of the restrictive provisions of the legal code.[37] A tradition gradually evolved according to which the Rana Prime Minister's word was regarded as above the law. In any case, copies of the legal code were not freely available to the public, and even government offices could secure a copy only through direct orders from the Rana Prime Minister.[38]

The Ranas' tendency to disregard the checks that the 1854 legal code sought to impose on their authority became pronounced particularly during the rule of Prime Minister Bir Shumshere (1885-1901). The preamble to the legal code, which had sought to circumscribe the authority of the King and the Prime Minister, was repealed in 1888.[39] Provisions which had given the legal code the status of constitutional law, as well as those which sought to confer on the civil and judicial administration a measure of autonomy vis-a-vis the political authority, shared a similar fate. The role of the legal code was, thereafter, limited to the fields of personal and administrative law.

The fate of the independent institutions created by Prime Minister Jung Bahadur to keep a watch on the administration was even worse. The anti-corruption court was abolished by Prime Minister Ranoddip Singh in 1878,[40] while the *Adalat Goswara* became merely an agency to scrutinize petitions submitted to the Rana Prime Minister.[41]

During the last decade of the nineteenth century, the bureaucracy too lost whatever autonomous character it had been able to acquire during the early years after the emergence of the Rana regime. In course of time, its aims, interests and orientations completely subserved those of the Rana rulers. The bureaucracy could neither secure an independent power and status base for itself, nor keep a "middle position" between the political elite and the common people.[42] Jung Bahadur's successors took a number of steps which served to narrow down the political horizon of the bureaucracy, and kept it oriented toward routine administrative tasks. For instance, payment of salaries was occasionally made conditional upon the full collection of revenue, and failure to make full collections was punished with a proportionate cut in salaries.[43] At times, officials were appointed on condition that they reduced administrative expenses by a specified amount.[44]

Moreover, the Rana system of administrative centralization had proceeded to ridiculous lengths by the last years of the nineteenth century. In early 1886, the chief of the elephant depot in Deukhuri asked for five rupees to buy ropes and other materials. His request was referred to the forest department, which approved it on the ground that "the government will incur losses if elephants escape for want of these materials." Finally, the case was represented before Prime Minister Bir Shumshere, who sanctioned the amount. Disbursement was then made by the revenue office of Banke district, located at Nepalgunj,[45] across a distance of at least two days' journey.

During the regime of Prime Minister Bir Shumshere, administrative policy became progressively revenue-oriented. In fact, the criterion for adopting any measure was whether or not it would increase revenue without inflicting undue hardships on the people.[46] Little attention was paid to measures for improving their general condition. For instance, repairs to a damaged irrigation channel in the Tarai region were, in one case, justified on the ground that the damage was causing loss to the government.[47] The potential increase in agricultural production and improvement in the condition of the peasantry that the irrigation facility would bring about were apparently not relevant factors influencing the decision to sanction funds for the project. Even the desire to avoid hardships to the people was motivated not by a sense of accountability for their welfare, but by the realization that it might be difficult to collect taxes from a dissatisfied peasantry. The Rana government was careful not to kill the goose that laid the golden eggs, but neither did it let the goose grow fat.

The Nature of the Rana Regime

It is not surprising, in these circumstances, that the Rana political system has been described as "an undisguised military despotism of the ruling faction within the Rana family over the King and the people of the country," under which the government functioned as an instrument to carry out the personal wishes and interests of the ruling Rana Prime Minister, its main domestic preoccupation being "the exploitation of the country's resources to enhance the personal wealth of the Rana ruler and his family."[48] Moreover:

As a system accountable neither to the King nor to the people, the Rana regime functioned as an autochthonous system, divorced from the needs of the people and even from the historical traditions of the country, and served only the interests of a handful of Ranas and their ubiquitous non-Rana adherents.[49]

But the goal of serving the interests of the handful of Ranas and their ubiquitous non-Rana adherents could not be pursued in isolation from political factors. The Ranas did not constitute a landed aristocracy which had succeeded in capturing political power. Rather, it was their political power that enabled them to acquire land and other economic resources. It was, therefore, a matter of supreme importance for them to establish and maintain a unified and centralized polity, and their own unchallenged authority over that polity, in place of the revenue farming-military complex that they had inherited from their Gorkhali predecessors.

The establishment of a unified and centralized polity made it necessary for the Ranas "to further the development of various types of free-floating, mobile resources not tied to any ascriptive groups and thus able to be freely accumulated and exchanged."[50] In other words, it was necessary for the Ranas to maximize the amount of monetary revenues that went directly to the central treasury, check the lavish grant or assignment of land and other economic resources to individuals, and extend state control over such economic resources as well as over sources of monetary revenue.

Limitations on the mobilization of resources in this manner were, however, imposed by the Ranas' own orientation, especially by their "identification with many ascriptive and traditional aspects of existing social institutions and values, as well as by their strong emphasis on their traditional legitimation."[51] For instance, the Ranas continued and even carried forward traditional land grant and assignment systems that checked the flow of revenue to the central treasury. Such a compromise with the goal of mobilizing "free-floating resources" was dictated by a two-fold need: to placate those groups that had traditionally benefitted from land grants and assignments, and to legitimize the Ranas' own acquisition of landed property through recourse to those traditional means. These limitations to the mobilization of resources were reinforced by the fact that some of the ascriptive and traditional elements in the society also provided the foundation of the civil administration that the Ranas created as one basis of their strength and the main medium for the execution of their policies. In fact, members of the old nobility, landlords, and other privileged groups in the society virtually monopolized the leading posts in the civil administration throughout the Rana period. The Rana government had, therefore, to be careful not to make a complete end of the landowning and other privileges traditionally enjoyed by these groups.

To sum up: Under the Rana political system, the Rana family held a monopoly in political power through which it controlled and exploited the nation's resources for its own benefit. However, in order to sustain that monopoly, the Ranas had to share the economic benefits of their rule with the aristocracy and the bureaucracy. Inasmuch as land was then the principal economic resource of the nation, the fundamental goal of the Rana political system was to garner the economic surplus generated by the peasantry for the benefit of the Rana ruler as well as the aristocracy and the bureaucracy, of which the Rana family itself was an important segment. This process of garnering the economic surplus generated by the Nepali peasant during the nineteenth century is by no means of mere historical interest. As a consequence of that process, individuals and groups who fulfilled no economic function were able to appropriate the major portion of what the peasant produced, whereas the peasant himself was left permanently stripped of capital. By letting the aristocracy and the bureaucracy share the benefits of their rule, the Ranas, no doubt, avoided an attack on their political authority, but only at the cost of the economic stagnation of the nation.

NOTES

[1]S. N. Eisenstadt, *The Political System of Empires*, New York, Free Press, 1963, p. 3.

[2]For an abstract translation of this Charter, see Satish Kumar, *Rana Polity in Nepal*, Bombay, Asia Publishing House, 1967, pp. 159-60.

[3]Barrington Moore, *Social Origins of Dictatorship and Democracy*, Middlesex, Penguin Books, 1969, p. 345.

[4]For the text of the "Treaty with Nipal Regarding the Restoration of the Western Tarai," which was signed on 1 November 1860, see Ramakant, *Indo-Nepalese Relations*, Delhi, S. Chand and Co, 1968, pp. 375-76.

[5]*Ibid*, pp. 78-97. This is how Ramakant characterizes the state of Nepal-India relations between 1818 and 1832.

[6]"Administrative Regulations for Morang District," Baisakh Badi 1, 1851 (April 1794) *RRC*, Sec. 3, Vol. 19, pp. 100-103; "Regulations for Territories under the Doti Administrative Headquarters Office," Kartik Sudi 8, 1908 (November 1851), *ibid.*, Sec. 24, Vol. 49, pp. 98-99.

[7]"Regulations for the Naya Muluk Region," Marga Badi 6, 1918 (November 1861), *ibid.*, Sec. 17, Vol. 47, p. 446.

[8]For the text of the treaty, see Ramakant, *op. cit.*, pp. 373-74. See also *ibid*, pp. 336-37.

[9]"Regulations for the Eastern Tarai Districts," Marga Badi 6, 1918 (November 1861), *RRC*, Secs. 2-11, Vol. 10, pp. 4-9; "Regulations for the Naya Muluk Region," Marga Badi 6, 1918 (November 1861), *ibid.*, Secs. 2-8, Vol. 47, pp. 440-43.

[10]Bhuwan Lal Joshi and Leo E. Rose, *Democratic Innovations in Nepal*, Berkeley and Los Angeles, University of California Press, 1966, p. 14.

[11]*Ibid.*, p. 40.

[12]T.B. Bottomore, *Elites and Society*, Middlesex, Penguin Books, 1971, p. 14.

[13]For abstract translations of the royal orders mentioned in this section, see Satish Kumar, *op. cit.*, pp. 158-60.

[14]"Rajkajko Ain" (State Affairs Act), Secs. 3-4. Information regarding the date when this law was first promulgated is not available, but its contents show that it was promulgated by Prime Minister Jung Bahadur. The earliest reference

to this law available to the author is contained in "Birta Land Grant to Prime Minister Ranoddip Singh," Ashadh Sudi 1, 1940 (June 1883), *RRC*, Vol. 32, p. 74. The *Raj Kaj Ko Ain* was first amended by Prime Minister Bir Shumshere in 1888.

[15]*Ibid.*, Sec. 21. In 1857, members of the Rana family were granted exemption from payment of the homestead tax. "Order regarding Serma-Tax Exemption for Members of the Rana Family," Bhadra Sudi 2, 1915 (August 1858), *ibid.*, Vol. 66, pp. 321-26.

[16]John Hicks, *A Theory of Economic History*, London, Oxford University Press, 1969, p. 18.

[17]Cf. "Appointment of Colonel Ripubhanjan Pande Chhetri to Discharge Revenue Collection and Judicial Functions in Morang District," Marga Badi 8, 1918 (November 1861), *RRC*, Vol. 10, pp. 189-91.

[18]"Audit Regulations for the Eastern Tarai Districts," Marga Badi 7, 1918 (November 1861), *ibid*, Sec. 14, Vol. 10, p. 245.

[19]"Survey Regulations for the Eastern Tarai Districts," Marga Badi 6, 1918 (November 1861), *ibid.*, Sec. 21, Vol. 10, p. 163.

[20]Mahesh C. Regmi, *A Study in Nepali Economic History: 1768-1864*, New Delhi, Manjusri Publishing House, 1971, p. 138.

[21]Satish Kumar, *op. cit.*, p. 103; *Regmi Research Series*, Year 6, No. 8. August 1 1975, pp. 150-53; "Order to Kharidar Devanarayan Upadhyaya[of the Kamandari Kitabkhana," Shrawan Sudi 12, 1906 (August 1949), *RRC*, Vol. 33, pp. 152-53.

[22]Cf., "Order regarding Appointments in Saptari Kathmahal Office," Poush Badi 11, 1942 (December 1885), *RRC*, Vol. 54, pp. 181-88.

[23]Hicks, *op. cit.*, p. 19.

[24]"Appointment of Bishnu Prasad Gurugharana Panditju as Judge," Falgun Badi 1, 1916 (February 1860), Marga Badi 11, 1917 (November 1860), and Magh Badi 1, 1919 (February 1863), *RRC*, Vol. 49, pp. 342-44, 352-54, and 369-70.

[25]Baburam Acharya, *Shri 5 Badamaharajadhiraja Prithvinarayan Shah*, Kathmandu, His Majesty's Principal Secretariat, Royal Palace, 2026 (1969), pt. 4, pp. 749-53; Satish Kumar, *op. cit.*, p. 102; "Order regarding Payment of Salaries from Revenues of Dafdarkhana Kumarichok," Poush Sudi 15, 1889 (January 1832), *RRC*, Vol. 27, p. 271.

[26]"Appointment of General Badrinarsingh Kunwar as Chief of Kumarichok," Marga Sudi 4, 1905 (November 1848*)*, *ibid.*, Vol. 33, pp. 79-82.

[27]Ministry of Law and Justice, "Rairakmko" (On Revenue Collection) in *Shri 5 Surendra Bikram Shahdevaka Shasankalma Baneko Muluki Ain* (Legal Code enacted during the reign of King Surendra Bikram Shah Dev), Kathmandu, the Ministry, 2022 *(1965)*, pp. 52-61.

[28]"Appointment of Subba Rudradatta Gurugharana Panditju as Chief of Dharmakachahari," Shrawan Badi 9, 1927 (July 1870*)*, in Yogi Naraharinath (ed.), *Sandhipatrasangraha* (A Collection of Treaties and Documents), Dang, the editor, 2022 (1966), pp. 143-44; *Regmi Research Series*, Year 7, No. 2, February 1, 1975, pp. 32-33, and Year 7, No. 7, July 1, 1975, p. 122.

[29]Ministry of Law and Justice, *Shri 5 Surendra . . .Muluki Ain*, Preamble, p. 2,

[30]Eisenstadt, *op. cit.*, p. 138.

[31]For instance, government officers were forbidden to assign private-owned tax-free lands to employees in lieu of their emoluments even on the orders of the King or the Prime Minister. The code prescribed that they would not be held guilty if they disobeyed such orders, but that obedience would be regarded as an act of guilt. "Jagga Jamin Goswarako" (on Miscellaneous Land Matters), in Government of Nepal, *Ain* (Legal Code), Kathmandu, Manoranjan Press, 1927 (1870), pt. 1, Sec. 19, pp. 16-17.

[32]Ministry of Law and Justice, "Adalatka Firyadi Ko" (On Complaints Filed in Courts) Sec. 2, in Ministry of Law and Justice; *Shri 5 Surendra . . . Muluki Ain*, p. 218.

[33]*Ibid.*, Preamble, p. 2.

[34]Ministry of Law and Justice, "Adalatiko" (on Judicial Matters), Sec. 19, in Ministry of Law and Justice, *Shri 5 Surendra. . . Muluki Ain*, p. 173.

[35]"On Complaints Filed in Courts," Secs. 20-21, pp. 222-23. (See n. 32).

[36]*Ibid.*, Sec. 10, p. 220; "On Judicial Matters," Sec. 21, pp. 171-172. (See n. 34).

[37]For instance, in 1874 Prime Minister Jung Bahadur sanctioned the conversion of the land assignment of a military officer into a cash salary "even though current law forbids such conversion." "Order regarding Conversion of Jagir Land Assignment of Lt. Janak Bahadur Adhikari Chhetri into a Cash Salary," Aswin Badi 4, 1931 (September 1874), *RRC*, Vol. 68, pp. 53-55. There are innumerable such cases in which Rana Prime Ministers wilfully acted in contravention of the law.

[38]In 1892, Lt. Colonel Sher Bahadur Thapa Chhetri of the Dailekh Administrative Headquarters Office reported to Kathmandu: "No copy of the new legal code has been received at this office. The local *Adalat* has refused permission to let us peruse or copy it, on the plea that it has received an order not to do so." The government then directed the *Ain Khana* (Law Department) to furnish to the Dailekh Administrative Headquarters Office a copy of those provisions only that did not deal with judicial administration. "Order regarding Supply of the Legal Code to the Dailekh Administrative Headquarters Office," Ashadh Sudi 1, 1948 (June 1892), *RRC*, Vol. 30, pp. 20-31.

[39]The 1888 edition of the legal code omits the preamble contained in the original edition. Government of Nepal, *Ain* (Legal Code), Kathmandu, Biradev Prakash Press, 1945 (1888), 5 pts.

[40]Satish Kumar, *op. cit.*, p. 104.

[41]Cf. "Order regarding Land-tax Assessments in Rautahat District," Magh Badi 7, 1942 (January 1886), *RRC*, Vol. 51, pp. 2025-40; "Appointment of Krishna Bahadur Shah and Narabhupala Shah as Mukhiya-Jimmawals in Musikot," Magh Badi 6, 1949 (January 1893), *ibid.*, Vol. 57, pp. 203-26.

[42]Eisenstadt, *op. cit.*, p. 171.

[43]Cf., "Order to Mukhiya Hiralal Singh Rajbhandari regarding Collection of Pota-tax in Kathmandu," Chaitra Badi 9, 1942 (March 1886), *RRC*, Vol. 40, pp. 40-45.

[44]Cf., "Order to Subba Devarika Lal Pande of the Bhaluhiya Birta Mal Office," Jestha Sudi 13, 1942 (June 1885), *ibid.*, Vol. 54, pp. 131-36.

[45]"Order to Captain Kirtiman Karki Chhetri of the Rapti-Deukhuri Dwar Office," Baisakh Sudi 7, 1943 (May 1886), *ibid.*, Vol. 56, pp. 215-19.

[46]Cf. "Order regarding Land-Tax Assessments in Rautahat District," Baisakh Sudi 4, 1942 (May 1885), *RRC,* Vol. 56, pp. 17-38.

[47]"Order regarding Construction of Irrigation Channels in Mahottari District," Jestha Badi 13, 1942 (May 1885), *ibid.*, Vol. 56, pp. 81-88.

[48]Joshi and Rose, *op. cit.*, p. 39.

[49]*Loc. cit.*

[50]Eisenstadt, *op. cit.*, p. 118.

[51]*Ibid.*, p. 120.

3

The Landowning Elite

Under the agrarian sytem that prevailed in Nepal during the nine-teenth-century, ownership of the land was normally vested in the state. For political and administrative reasons, large areas of state-owned lands were granted on freehold tenure to members of the aristocracy and the bureaucracy, religious and charitable institutions, etc. The actual cultivator, therefore, usually held his land on ten-ancy. He paid rent either to the government or to individuals or institutions who were beneficiaries of state land grants. The fruits of cultivation were thus traditionally divided into two parts: *talsing-boti*, or the landlord's share, and *mohi-boti*, the portion of the crops that the cultivator was allowed to retain for himself. In effect, the system meant that the surplus produce of the land belonged to aristocratic and bureaucratic groups in the society, whereas the peasant was a mere instrument to work the land and produce taxes for their benefit.

Traditional Lard Tenure Systems[1]

It would be appropriate to begin with a description of the tradi-tional relationship between the state and land in Nepal, inasmuch as that relationship has determined not only the nature of the groups that enjoyed the privilege of appropriating surplus agricul-tural production, but also the *modus operandi* of such appropriation. The key element in that relationship is the state's ownership of land and other natural resources in its territories. Both by law and by tradition, land was the property of the state. As both sovereign and landowner, the state was entitled to the payment of a part of the produce of the land as tax or rent. This system of state land-ownership is traditionally called *raikar* in Nepal.

The institutional character of state ownership of the land was circumscribed by the existence of a communal form of land tenure, known as *kipat*, in some parts of the hill regions. Under the *kipat* system, land belonged to the local ethnic community under customary law, not to the state under the statutory law. Several Mongoloid or other autochthonous communities in the hill regions had been able to retain their customary occupation of lands on a communal basis under the *kipat* system, although the lands were divided into individual holdings for purposes of actual cultivation. *Kipat* was a form of communal landownership, under which each person had the right to the exclusive use of a particular piece of land. However, his rights to dispose of the land were restricted on the theory that the land belonged to the community as a whole. The government, therefore, had no power to impose taxes and rents on *kipat* lands; it only exercised its sovereign power of taxation of individual *kipat* owners. *Kipat* lands were owned by tribal communities of *kirat* ethnic stock mainly in the far-eastern hill region beyond the Dudhkoshi river, who controlled almost the entire area of both rice and hill-side lands. They were able to retain this control in substantial measure within a broad framework of local autonomy which was a condition of their incorporation into the Gorkhali empire during the mid-1770s. There were scattered settlements of *kipat*-owning communities in other parts of the hill region also, including Kathmandu Valley.

Occasionally, the state alienated its ownership rights in *raikar* land, as well as its sovereign power of taxation, to individuals. These grants were made under three different systems: *rajya*, *birta* and *jagir*. These systems formed the foundation of the political and administrative set-up introduced after the political unification of the Kingdom during the latter part of the eighteenth century. Only *raikar* lands, which belonged to the state, were so granted. Because the state did not own *kipat* lands, it could not grant those lands as *rajya* or *birta*; only the taxes collected from individual *kipat*-owners were often included in *jagir* assignments.

The Rajya, Birta and Jagir Systems

The term *raja* literally means a king, and *rajya*, a kingdom. In post-1770 Nepali historiography, however, *rajya* means a vassal principality in the Gorkhali empire, which usually enjoyed a substantial measure of local autonomy. The *rajya* system emerged in the course of the political unification of Nepal during the latter part of the eighteenth century. A few principalities in the far-western hill region between the Bheri and Karnali rivers were then incorporated into the Gorkhali empire by diplomacy, rather than by conquest. The erstwhile ruler was permitted to retain his authority with some measure of autonomy in internal administration, subject to the general suzerainty of Kathmandu. Such a policy made it possible for the Gorkhali rulers to achieve the political unification of the country with a minimum of military and administrative effort. At the time of the commencement of Rana rule in 1846, these vassal principalities included Bajura, Bajhang, Darna, Jajarkot, Salyan, Mustang and Udayapur.

Birtas were given to individuals in appreciation of their services to the state, as ritual gifts, or as a mark of patronage. The recipients included priests, religious teachers, soldiers, members of the nobility, and the royal family. Such grants had been made in Nepal at least since the fifth century. Often lands were endowed by the king, or any individual, for temples and monasteries, as well as for other religious and charitable purposes. Lands endowed in this manner were known as *guthi*. In the present context, we shall use the term *birta* to include *guthis* also, inasmuch as the tenurial and fiscal privileges attached to both categories of land grants were more or less identical. *Jagir* lands, on the other hand, were assigned to civil or military employees and government functionaries of different categories as their emoluments. The government of Nepal traditionally paid its employees cash salaries only if lands suitable for assignment as *jagir* were not available. The *jagir* system performed two essential functions: it lessened the cost of the military establishment, and appeased the land-hunger of an army that was composed, for the most part, of peasants.

Although the state alienated its landownership and sovereign rights in the land under the *rajya*, *birta* and *jagir* systems, the terms and conditions of alienation were different in each case. *Rajyas* and *jagirs* were made on an exclusively usufructuary basis; the recipients were not permitted to transfer or sub-divide the lands mentioned in the grant. Government employees in Nepal were traditionally appointed only for a one-year term, hence all *jagir* lands reverted to the state at the end of this term. *Birta* alone was a form of private property which usually could be subdivided, inherited, sold, mortgaged or bequeathed.

Notwithstanding these variations in the nature of landownership under the *rajya*, *birta* and *jagir* systems, the rights and privileges that the beneficiaries enjoyed were more or less identical. For the purpose of the present study, these rights and privileges may be enumerated as follows: The right to a share in the produce of the land; the right to the proceeds of miscellaneous taxes and levies collected from the inhabitants of the lands and villages granted; the right to exact unpaid labour on a compulsory basis from those inhabitants, and the right to dispense justice.

These rights and privileges that *rajas*, *birta*-owners, and *jagirdars* enjoyed in nineteenth-century Nepal will be discussed in greater detail at appropriate places in the following chapters. So far as judicial authority is concerned, it was essentially secular in character. That is to say, it was concerned with the enforcement of civil and criminal law, or rules which govern the relations of men in civil society. Enforcement of Hindu customs and sanctions that provided for ritual purity, or of customary rules of sexual and commensal intercourse among members of communities that had not yet been incorporated into the traditional four-caste system of the Hindus, lay beyond the purview of secular justice. Secular justice was administered at two levels: central and local. Central-level justice was limited to crimes of a serious nature, collectively known as *Panchakhat*.[2] Punishment for *Panchakhat* crimes was usually inflicted directly from Kathmandu, although a few *rajas* also enjoyed that ¡authority[3] before 1889.[4] The power to administer justice in minor cases other than *Panchakhat* usually devolved on local functionaries. When lands and villages were granted to individuals under the *rajya*, *birta*, and *jagir* systems, the *raja*, *birta*-owner, or *jagirdar* became the beneficiary of such devolution. For most of the cases that would come up in day-to-day life, therefore, the peasant lived under the control of his lord.

The powers of the government with regard to lands and villages alienated through *rajya*, *birta*, and *jagir* grants did not include the powers of taxation and police power. The inhabitants of such areas had no direct contacts with the central government in the ordinary affairs of life. The modern political concept of the state exercising full sovereign authority over all areas and all classes of people in the territories comprising its dominions was hence unknown in nineteenth-century Nepal. The economic power of the government of Nepal, and hence its political power, was eroded by the authority granted to the landowning elite. From the viewpoint of the peasantry, the authority of the landowning elite replaced the state authority. The authority that the government enjoyed over lands and subjects in the territories comprising its dominions was, therefore, residual, and continually encountered a maze of private jurisdictions.

Such fragmentation of authority did not mean, however, that nineteenth-century Nepal was divided into two watertight compartments under the jurisdictions of the government and the landowning elite. The inhabitants of *rajya, birta* and *jagir* lands and villages owed allegiance to the *raja, birta*-owner, or *jagirdar* in matters concerning payment of agricultural rents, taxes, labour services, and the settlement of their disputes, but there were several other spheres in which they functioned under the direct jurisdiction of the government. For instance, they were under obligation to provide unpaid labour to the government for public purposes. People were recruited directly to the army without any reference to their *raja, birta*-owner, or *jagirdar*. Moreover, no distinction was made between the inhabitants of *raikar* lands on the one hand and *rajya, birta*, and *jagir* lands on the other as regards the jurisdiction of the government in the dispensation of justice in *Panchakhat* crimes. Finally, any subject, irrespective of his wealth and status, was free to approach the King for the redress of his grievances against the injustice or oppression committed by any member of the nobility, government servant, or local tax-collecting functionary.[5]

The *birta* and *jagir* systems were used in nineteenth-century Nepal as a means for the enrichment of the political elite whose power was based on control of the administration rather than on the ownership and inheritance of property.[6] An example may be cited to illustrate the manner in which the Gorkhali nobility used these systems to enrich itself. Mathbar Singh Thapa, a nephew of Prime Minister Bhimsen Thapa (1806-37), fled to India as a penniless refugee when his uncle was dismissed from office and committed suicide in jail. His *birta* lands and other property were confiscated. He was recalled by King Rajendra Bikram and appointed Prime Minister in April 1839. His old *birta* lands were restored, and he was granted approximately 2,200 *ropanis* of rice-lands and revenues amounting to Rs 10,120 as his *jagir* emoluments. During his Prime Ministership of a little more than two years, Mathbar Singh Thapa obtained from the King fresh *birta* grants for as much as 25,346 *bighas* of land in the rich agricultural districts of Bara, Rautɑhat, Saptari, and Mahottari in the eastern Tarai region, and cash revenues amounting to Rs 10,562 from villages in the hill regions adjoining Kathmandu Valley. These personal grants were in addition to those made in his name for endowment as *guthi* to temples in different parts of the country.[7] The point that needs to be stressed in this context is that it was his political power that alone made it possible for Mathbar Singh Thapa to acquire such vast landed property within a period of less than two years. The entire property was confiscated after he was assassinated in May 1845.

Landowning Elites and Village Landlords

Inasmuch as *rajas*, *birta*-owners and *jagirdars* derived their political and economic rights and authority from the state, they were able to combine political control with economic exploitation. They did not owe their status and privileges to any economic services rendered to the local agrarian community, such as the supply of credit or of agricultural implements. They enjoyed the ascriptive right to collect a part of the peasant's produce in the form of rents, but were not liable to provide any compensatory services or benefits. Their high social status, as well as their political, military and administrative responsibilities, also made personal cultivation out of the question. *Rajas*, large *birta*-owners, and *jagirdars* were, therefore, absentee landlords, who usually did not live in the villages where their lands were situated, and played neither a direct nor a supervisory role in agricultural production.

In particular, cultivation of *jagir* lands by *jagirdars* themselves was seldom practicable.[8] The *jagirdar's* tenure was uncertain and often he was dismissed after he had already sown his fields. In that event, he lost not only the expenses he had incurred in sowing his fields, but even found himself suddenly without a roof over his head. Additional problems were created by the increase in the number of *jagir* land assignments consequent to the increased scale of military recruitment. It was then seldom possible to allot lands of convenient location to *jagirdars*. Because of these problems, *jagirs* eventually became a mere assignment of rents, which the *jagirdar* was permitted to collect from the cultivator only on the authority of certificates, known as *tirjas*, issued for each crop every year by the central government.[9] *Jagir* landownership was thus effectively divorced from actual cultivation of the land during the early years of the nineteenth century.

In the present study, the term landowning elite will be used to describe *rajas*, large *birta*-owners, and *jagirdars*. The term excludes small *birta*-owners, who lived in the village where their *birta* lands were situated, as well as *zamindars* and other categories of local landholding groups whose income consisted of the difference between the taxes they paid to the government and the rents they collected from the cultivator. These formed a group which may be described as village landlords.

The basis for the differentiation between the landowning elites and village landlords can be easily explained. *Rajas*, large *birta*-owners and *jagirdars* were usually absentee rent-receivers, whereas village landlords lived in the village and played an essential role in agricultural production. Although village landlords were usually able to accumulate resources in excess of their actual consumption needs, available evidence suggests that they used this surplus primarily for moneylending, acquisition of lands, often through the foreclosure of mortgages, financing of land reclamation and irrigation projects, and trade in agricultural produce. Village landlords were thus an important source of credit supply and capital investment in the village. They comprised a part of the local community whose needs and problems coloured their relations with the outside world in substantial measure. In contradistinction, the interest of the non-resident landowning elites was confined to the amount of income they could collect from the lands granted to them by the state.

The distinction between the landowning elite and village landlords in nineteenth-century Nepal will be illustrated by citing the examples of Prime Minister Mathbar Singh Thapa and the poet Bhanubhakta Acharya (1814-69). An earlier section had referred to the large areas of *birta* and *jagir* lands in different areas of the the country that Mathbar Singh Thapa had acquired during his tenure as Prime Minister. The lands and villages comprising these *birtas* and *jagirs*, and their inhabitants, were of interest to him solely as a source of income. It is indeed doubtful whether Mathbar Singh Thapa ever set foot on these lands and villages, or provided any assistance in cultivation. Income from these *birtas* and *jagirs* was not spent in the villages where it was collected. For the peasantry, therefore, Mathbar Singh Thapa was nothing more than a parasitic landlord.

In contradistinction, Bhanubhakta Acharya, who lived in the village of Ramgha of Tanahu district in the western hill region, belonged to the class that we have described as village landlords. His lands apparently fetched him an adequate income, and even left him enough to function as a moneylender.[10] Bhanubhakata Acharya had an allotment of *jagir* lands in his village, and often undertook land reclamation and irrigation schemes in cooperation with local landholders.[11] Trade in agricultural produce through the collection of *jagir* rents on brokerage was a side occupation, and his caste status and education permitted him to function as a priest and astrologer as well.[12] To be sure, both Mathbar Singh Thapa and Bhanubhakta Acharya lived mainly on the economic surplus that they collected from the peasantry, but Mathbar Singh Thapa enjoyed the right to a share in the peasant's produce through political power, whereas Bhanubhakta Acharya achieved his position by economic means within the existing framework of land tenure and taxation. Their contrasting roles in the agrarian system are reflected by the fact that Mathbar Singh Thapa provided no commensurate services to the local agrarian community, whereas Bhanubhakta Acharya functioned as a source of credit and capital investment in the village.

The present study will deal with the ascriptive right that the state granted to *rajas*, *birta*-owners, and *jagirdars* to partake of a share in the fruits of the peasant's labour, not the rights of village landlords. The peasant was compelled to relinquish the major portion of the output to the landowning elites because of the political control they wielded over state apparatus. However, these groups did not provide any political or economic benefits to the peasant. Unlike the feudal tenant in medieval Europe, the Nepali peasant in the nineteenth century was not "protected, defended and warranted"[13] by the *raja*, *birta*-owner or *jagirdar* whose lands he tilled. It is to the fundamental fact of exploitation that the agrarian system of Nepal during the nineteenth century owes many of its outstanding characteristics.

Rana Policies

The basic nature of the *rajya*, *birta* and *jagir* systems, as des-bed in the foregoing sections, remained more or less unchanged after the advent of Rana rule in 1846. The Ranas retained these systems in order to mobilize political support for their rule and finance the civil administration and the army despite a chronic shortage of cash for the payment of salaries. Their primary objective was to establish a centralized and unified polity in which political power was monopolized by the Rana family. Land grants under the *rajya*, *birta* and *jagir* systems helped the Ranas to extend their political control over elite groups in the society in pursuit of that objective.

Two consequences followed from the Ranas' manipulation of the *rajya*, *birta* and *jagir* systems to entrench their control over the political classes as well as over the civil and military establishment of the Kingdom. One was a change in the composition of the elite groups who owned land under these systems. Many leading members of the traditional nobility were massacred or banished from the kingdom during the Rana takeover of power in 1846. Their *birta* and *jagir* lands were confiscated by the government and re-allotted to those members of the nobility who supported the Ranas.[14] Fresh grants under the *rajya*, *birta* and *jagir* system naturally became the monopoly of the Rana powers-that-be.

A mere redistribution of the benefits of ascriptive landownership, however, was of little consequence from the viewpoint of the peasant. Irrespective of who owned the land he tilled, his fiscal and other obligations remained unaffected. The second consequence of the Ranas' land policy, an increase in the area covered by the *rajya*, *birta* and *jagir* systems, was, therefore, of greater importance. It meant that the state was deprived of the revenues of progressively large areas of lands, and that an increasing number of freehold cultivators or peasants, who paid no taxes to the state, came under the personal authority and control of *rajas*, *birta*-owner, and *jagirdars*. The following sections will discuss in detail the manner in which the area covered by these systems increased after the advent of Rana rule.

(a) Rajya Policy

It has been mentioned before that during the latter part of the eighteenth century, several independent principalities in the western hill region were incorporated into the Gorkhali empire and recognized as autonomous *rajyas*. The number of *rajyas* increased nearly threefold during the period of Rana rule. Before 1846, there were only seven *rajyas* in Nepal; by the end of the nineteenth century, at least 13 more *rajyas* had been added to the list: Achham, Bhirkot, Dhurkot, Dullu, Galkot, Garhunkot, Jumla, Kaski and Lamjung, Malbara, Malneta, Musikot, Khumrikot, and Pyuthan.

Rana policy toward the *rajya* system was initially dictated by the need to win over the support of the influential class of *rajas* for the new regime. The allegiance of a political group can be won over only when it is convinced either that offering such allegiance will benefit it economically or politically, or that withholding of allegiance will lead to the loss of existing economic or political privileges, or both. Those *rajas* who aligned themselves with the new regime were rewarded with land grants, cash allowances, and other perquisites,[15] whereas those who did not, lost their *rajyas*. For instance, the *raja* of Baghang was accused of conspiring against Prime Minister Jung Bahadur and so was deposed and replaced by his son, Bhupendra Singh, in early 1850.[16] The *raja* of Bajure was similarly disposessed of his territories in 1856; only in 1879 did Prime Minister Ranoddip Singh restore the *rajya*.[17]

To be sure, not all these newly-created *rajyas* enjoyed the same status. The *rajyas* recognized by the Gorkhali rulers before the emergence of Rana rule enjoyed internal autonomy, subject only to the payment of a nominal annual tribute to the royal court in Kathmandu. In 1767, for instance, the *raja* of Jajarkot was granted full internal autonomy subject to the payment of an annual tribute of Rs 701.[18] The *raja* of Bajura was accorded a similar status in 1791 on payment of Rs 500 every year.[19] The *rajas* created by the Ranas, on the other hand, enjoyed somewhat less authority. Most of them were granted the privilege of collecting revenues in their old territories on a contractual basis on behalf of the government or of individual *jagirdars*, while a few were given the title of *raja* on a purely honorific basis.

The new *rajyas* created by the Ranas during the latter part of the nineteenth century on the condition that the *rajas* collected revenues in their territories on a contractual basis included Achham,[20] Dullu,[21] and Garhunkot.[22] A description of the functions and privileges of the *raja* of Achham, as mentioned in a royal grant made in 1887,[23] will illustrate the nature of *rajyas* of this category. Dal Bahadur Shah, a prince of the royal house of Achham, was granted a revenue collection contract in that year in the territories by the old principality of that name, as well as the title of *raja* on his marriage to a grand-daughter of Prime Minister Ranoddip Singh. His sole obligation was to collect taxes on *jagir* lands at the prescribed rates in the territories of the *rajya* and transmit the proceeds to the appropriate *jagirdars*. Otherwise, he was granted full ownership rights over uncultivated lands in the areas under his jurisdiction, and authority to collect and appropriate the proceeds of customary taxes and levies which had not been included in *jagir* assignments. The *raja* was also granted authority to dispense justice, except in *Panchakhat* crimes.[24]

The policy of granting the title of *raja* on an honorific basis, without assigning territories, was initiated by Prime Minister Jung Bahadur. He solicited financial and military assistance from several princely families that had been deposed during the Gorkhali conquests for the 1855-56 Nepal-Tibet war and the Nepali military expedition to India during the 1857 rebellion, possibly with the objective of testing their loyalty to his regime. The response was positive and unanimous. Subsequently, some of these princes were recognized as *rajas* in appreciation of their services during these campaigns, but only on an honorific basis. They were not awarded the status of autonomous rulers in their old territories. Many of the new *rajas*, in fact, were only given some *birta* lands in their old territories. In 1868, for instance, Prithvi Bam Malla, a prince of the displaced royal house of Galkot, was granted *birta* lands fetching an income of Rs 300 a year and recognized as a *raja*.[25] The *raja* of Bhirkot[26] was also a *birta*-owner without any pretensions to the autonomy enjoyed by the *rajas* of the other two categories.

The powers and privileges attached to the *rajya* of Kaski and Lamjung belong to a separate category. These two abolished *rajyas* were restored in 1856, and granted not to the direct heirs of their old ruling families but to Prime Minister Jung Bahadur, with the title of Maharaja.[27] Eventually, the posts of Maharaja of Kaski and Lamjung and Prime Minister were combined in the seniormost member of the Rana family.

(b) Birta Policy

The Rana period also witnessed a steady expansion of the area under *birta* tenure. The Rana rulers retained the traditional practice of making *birta* land grants to win over the loyalty of Brahmans, leading members of the nobility, and other politically influential groups. The beneficiaries included widows and dependents of members of the Rana family and other high-ranking persons, including members of the families of those nobles who had gone into exile, and had subsequently received pardon and returned to Nepal. In addition, the Ranas utilized their political power to enrich their own family through extensive *birta* grants. This two-fold objective was pursued with such vigour that in 1950, when Rana rule came to an end, at least one-third of the total culti-vated area in the Kingdom was under *birta* tenure, and approxi-mately three-fourths of the total area under *birta* tenure belonged to members of the Rana family.[28]

There was usually one important difference in the nature of *birta* grants made to members of the Rana family and other individuals. Most grants made to members of the Rana family were inheritable and unconditional. They entitled the beneficiaries to exercise all the traditional rights and privileges of *birta* landownership in full. On the other hand, most *birta* grants made to non-Ranas were condi-tional and carried only a part of these rights and privileges. For instance, these grants were often non-inheritable, or even taxable.[29]

The Rana family, as mentioned above, was the major beneficiary of *birta* land grants made after 1846. In 1861, it obtained the big-gest *birta* grant ever made in Nepal, comprising the present dis-tricts of Banke, Bardiya, Kailali, and Kanchanpur in the far-western Tarai region.[30] Nepal had lost these territories after the 1814-16 Nepal-British war. The British Indian government restored them in 1860 in appreciation of the military assistance rendered by Nepal during the 1857 rebellion in India. King Surendra then granted half of these territories as *birta* to Prime Minister Jung Bahadur, and the other half to his six brothers.[31]

Some time during the 1860s, legislation was enacted according to which no Prime Minister was permitted to accept *birta* grants from the King, except in territories added to the Kingdon during his term of office.[32] This measure was obviously intended to legitimize the above-mentioned grant. It also banned the grant of cultivated lands in the old territories as *birta* to members of the Rana family, possibly because such grants would cause a loss of revenue to the government, although *birta* grants on forest lands in these territo-ries were permitted.

These statutory limitations on the acquisition of *birta* lands by the Prime Minister and other members of the Rana family were seldom actually observed. In 1861, for instance, several villages in Nuwakot district were granted to Prime Minister Jung Bahadur as *birta* in 1861 in appreciation of the services he had rendered in escorting King Surendra during a pilgrimage.[33] Possibly to avoid any impression of illegality, during the later years of his rule, Jung Bahadur often acquired *birta* lands only after paying the price to the government. It was in this manner that in 1873 he obtained a *birta* grant of approximately 250 *bighas* in Sarlahi district on payment of Rs 125,000 in Indian currency to the government.[34] There is evidence that Jung Bahadur's successors were less scrupulous in ensuring the legality of their *birta*-land acquisitions.[35]

Nevertheless, not all *birta* grants made by the Ranas were intended to enrich themselves and their supporters. The *birta* policy of the Ranas had other objectives also, such as promoting land reclamation. Most *birta* grants made in the Tarai during the early Rana period, in fact, appear to have been made on waste or virgin lands,[36] the obvious objective being to promote the development of that region. In line with that policy, legislation was enacted in 1854 permitting any person who reclaimed forest lands to use one-tenth of the reclaimed areas as *birta*.[37] In other words, the government granted him tax exemption for the entire reclaimed area for five years, an⌟ one-tenth of this area on a tax-free basis in perpetuity. Even persons holding top-ranking positions in the civil service or the army found these grants attractive,[38] presumably because their tenure on the *jagir* lands assigned to them was precarious and uncertain. The possibility of reclaiming uncultivated lands in the Tarai accordingly guaranteed them a measure of financial security, which was not attainable otherwise.

(c) Jagir Policy

Like their predecessors, the early Rana rulers acted on the principle that government employees should not be paid cash salaries so long as lands were available for assignment as *jagir*.[39] The government preferred to remunerate its employees through assignments of lands, because it was chronically short of cash.[40] The land policy followed during the eighteenth and nineteenth centuries consequently led to a decline in the area under *raikar* tenure.

A *jagirdar* could request the government at any time that his cash salary be replaced wholly or partly by a *jagir* land assignment. His request was usually granted if lands were available.[41] At times, however, the shortage of lands left no alternative for the government but to pay salaries in cash.[42] Such shortage, no doubt, checked the expansion of the *jagir* system, but it also highlights the manner in which agricultural lands in nineteenth-century Nepal were used for the benefit of select groups in the society, rather than as a source of public revenue.

The State and the Landowning Elite

As a result of the policies followed by the Rana rulers, the area covered by *rajya, birta* and *jagir* grants increased considerably during the second half of the nineteenth century. In fact, almost the entire cultivated area in the Kingdom, particularly in the hill region, had been alienated by the state under these systems.[43] The income from lands so alienated accrued to *rajas, birta*-owner, and *jagirdars*, not to the state exchequer. The amount of income that these landowning groups appropriated from the land was consequently much more than what the government itself was able to collect. In 1853, for instance, the total cash receipts of the government from all sources, including land, amounted to a meager Rs 926,273 against an assessed figure of Rs 1.96 million, whereas the official value of *jagir* land assignments amounted to Rs 1.92 million. In 1857, land and forest revenue from the eastern Tarai region amounted to Rs 749,836, whereas owners of *birta* lands in that region were collecting an income of at least Rs 880, 296.[44] *Rajya, birta* and *jagir* land grants throughout the Kingdom consequently absorbed the lion's share of the resources that were actually collected from the peasantry. On the other hand, the central treasury in Kathmandu faced a chronic shortage of cash.

As noted in the previous chapter, the mobilization of free-floating monetary resources was one of the general goals of the Rana regime. Land grants under the *rajya, birta* and *jagir* systems naturally hindered efforts aimed at such mobilization. Accordingly, even while making *rajya, birta* and *jagir* grants with social, political, and administrative objectives, the Rana rulers sought to minimize the loss of revenue that those grants entailed. At the same time, measures aimed at maximizing the monetary revenues of the government were sufficiently adroit to leave more or less untouched the basic nature and privileges of ascriptive landownership under the *rajya, birta* and *jagir* systems.

The Rana government, therefore, initiated a series of measures

aimed at lessening the impact of the *rajya, birta,* and *jagir* systems
on national finance, albeit without affecting the basic structure of
these systems. The first of these measures was the abolition of the
iagir system in the Tarai region in 1853.[45] The decision constituted
part and parcel of efforts to strike a balance between the alienation of revenue through land grants and the ascriptive privileges of
the landowning elites. The Rana government made an attempt to
justify that decision by pointing out that "because lands in the
Tarai region have been assigned as *jagir* to civil and military officials, the amount of revenue deposited in the treasury is meager."[46]
It should be noted, at the same time, that the measure did not
reduce the total area under *jagir* tenure, inasmuch as the *jagirdars*
who lost their lands in the Tarai obtained replacements from Crown
lands in the hill regions.[47] It is also significant that the Rana rulers'
desire to augment revenue from the Tarai did not have any effect
on the practice of making *birta* grants in that region.

The Rana government also abolished a number of facilities and
privileges traditionally enjoyed by *birta*-owners and *jagirdars,* thereby diverting new sources of revenue to the state treasury. Several
special levies which had been included in *jagir* land assignments were
withdrawn.[48] The practice of including revenue from fines, customs
and transit duties in *jagir* assignments was banned.[49] Forests on
jagir were taken over by the state,[50] and *birta*-owners in the Tarai
were obligated to sell timber from their forest lands to the government at statutory prices.[51] *Jagidars* were thereafter left solely with
the income accruing from agricultural rents and homestead taxes.
Limitations were imposed also on the exercise of judicial authority
by *birta*-owners and *jagirdars.* The 1854 legal code precisely defined
their jurisdiction: they were not permitted to adjudicate in cases
which involved a claim of more than Rs 500 each.[51] The objective
of this measure was to expand the jurisdiction and hence the income
of the government courts that the Rana rulers organized in different
parts of the country. Government courts were empowered to hear
appeals against decisions made by *birta*-owners and *jagirdars.*

The polices followed by the Rana government thus led to a significant increase in the total area alienated by the State under the
rajya, birta and *jagir* systems. Measures were, no doubt, taken
from time to time to bring *rajas, birta*-owners and *jagirdars* under
the tighter control of the central government. Thanks to these measures, these landlords probably wielded less police and adminis-

trative authority over the peasantry at the end of the nineteenth century than at the time of the commencement of the Rana rule. However, these measures had little impact on the basic right of the landowning groups to appropriate a share of the peasant's produce without providing any commensurate service in return. In other words, *rajas*, *birta*-owners, and *jagirdars* continued to occupy the role of parasitic landlords whose income from rents was available neither for investment in agricultural development nor for mobilization as the tax revenues of the state.

NOTES

[1]This section is based on Mahesh C. Regmi, *Land Tenure and Taxation in Nepal*, Berkeley, University of California Press, 1963-68, 4 vols, and *Land Ownership in Nepal*, Berkeley, Los Angles, and London, University of California Press, 1976, pp. 15-21.

[2]The term *Panchakhat*, during the eighteenth and nineteenth centuries, denoted the crimes of bribery, smuggling, murder (including infanticide), assault resulting in the shedding of blood, and cow-slaughter. ("Appointment of Chautariya Dalamardan Shah as Ijaradar in Bara and Parsa," Ashadh Sudi 6, 1843 (June 1786), *RRC*, Vol. 25, p. 146.) These crimes were punishable through "confiscation of the whole estate, banishment of the whole family, degradation of the whole family by delivering the members to the lowest tribes, maiming the limbs, and death by cutting the throat." Francis Hamilton, *An Account of the Kingdom of Nepal*, (reprint of 1819 ed.), Delhi, Manjusri Publishing House, 1971, p. 103.

[3]These *rajyas* included Jajarkot ("Confirmation of Gajendra Shah as Raja of Jajarkot," Magh Sudi 5, 1825 [February 1769], in Yogi Naraharinath [ed.], *Sandhipatrasangraha* [A Collection of Treaties and Documents], Dang, the editor, 2022 [1965], p. 4); Bajura ("Reconfirmation of Narendra Bahadur Shah as Raja of Bajura," Jestha Badi 5, 1944 [May 1887), *RRC*, Vol. 50, pp. 16-20); and Achham ("Confirmation of Dal Bahadur Shah as Raja of Achham," Poush Badi 11, 1934 [December 1887], *ibid.*, Vol. 55, pp. 681-88).

[4]On Kartik Sudi 8, 1946 (November 1889), all previous royal charters relating to the recognition and reconfirmation of *rajyas* in the Kingdom were withdrawn, and fresh charters were issued. One of the provisions of the new charters was that *rajas* should not adjudicate in *Panchakhat* cases in the future. The new provision is cited in: "Replacement of Bikram Bahadur Singh by his son, Jayaprithvi Bahadur Singh, as Raja of Bajhng," Shrawan Badi 1, 1947 (July 1880), *ibid.*, Vol. 50, pp. 448-55.

[5]In 1812, for instance, Birabhadra Kunwar was appointed as an officer at

the royal palace and provided, *inter alia*, with the following authority: "In case any person, high or low, rich or poor, and belonging to any of the four castes and thirty-six sub-castes, approaches the palace with the complaint that he has been subjected to highhandedness or injustice by any royal priest, member of the nobility, or civil or military officer, soldier, or tax-collection official or functionary in any part of our Kingdom, listen to his complaint attentively and ascertain what he wants. Refer the matter to us along with your recommendation, and take action as ordered by us." ("Regulations in the Name of Birabhadra Kunwar," Magh Sudi 5, 1868 [February 1812], *RRC*, Secs. 1-2, Vol. 40, pp. 358-60).

[6]T.B. Bottomore, *Elites and Society*, Middlesex, Penguin Books, 1971, p. 44.

[7]"Birta Lands of Prime Minister Mathbar Singh Thapa," *Regmi Research Series*, Year 8, No. 3, March 1, 1976, pp. 46-49.

[8]According to Kirkpatrick: "Though one or two men might, if time allowed, plough a kaith (i.e., one *khet*, or 100 *muris* of rice lands) very well, yet as several hands are required to sow and reap it, there can be but a small proportion of it separately managed by the Jaghiredar, or occupier, however numerous his family may be. He generally, therefore, satisfies himself with cultivating a few Moories, in which he usually raises fruit or vegetables, leaving the rest to the Mohi, or undertaker, with whom he engages." (William Kirkpatrick, *An Account of the Kingdom of Nepaul*, [reprint of 1811 ed.] New Delhi, Manjusri Publishing House, 1969, p. 98). However, *Jagirdars* were themselves responsible for necessary arrangements to have their *jagir* lands cultivated. In 1796, for instance, all categories of *jagirdars* belonging to the army were told that "you shall not be entitled to any rents if you cannot cultivate your [*jagir*] lands and let them remain uncultivated." ("Order regarding Jagir Land Assignments to the Army." Ashadh Sudi 2, 1853 [June 1796], *RRC*, Vol. 20, pp.48-50.)

[9]Cf., "Order regarding Land Reclamation and Rent Collection in Majhkirat," Jestha Badi 9, 1880 (May 1823), *ibid.*, Vol. 43, pp. 440-42; "Arrangements Regarding Collection of Rents on Jagir Lands of Jwaladal Company," Ashadh Badi 13, 1882 (June 1825), *ibid.*, Vol. 34, p. 14.

[10]"Personal Bond Signed by Sadhuram," Falgun Badi 12, 1918 (February 1862), in Balakrishna Pokhrel [ed.], *Pancha Saya Varsha* (Five hundred years of Nepali Prose Literature), Lalitpur, Jagadamba Prakashan, 2020 (1963), pp. 570-71.

[11]"Order to Gorkha Adalat regarding Complaint against Bhanubhakta Pandit and Others," Kartik Badi 14, 1921 (October 1864), *RRC*, Vol. 49, pp. 402-4.

[12]Baburam Acharya, *Purana Kavi ra Kavita* (Old Poets and their Poetry), Kathmandu, Sajha Prakashan, 2023 (1966), p. 110.

[13]Marc Bloch, *Feudal Society*, London, Routledge and Kegan Paul Ltd, 1975, Vol. 1, p. 246.

[14]Cf., "Confirmation of Grant of House in Kathmandu, to Lt. Randhwaj Adhikari Chhetri," Marga Badi 4, 1921 (November 1864), *RRC*, Vol. 47, pp. 124-27.

[15]Cf., "Petition of Raja Bikram Bahadur Singh of Bajhang," Bhadra Sudi 11, 1931 (September 1874), *ibid.*, Vol. 32, pp. 778-79; "Order regarding Per-

quisites of Raja Dip Narayan Shah of Jajarkot," Ashadh Sudi 14, 1911 (July 1854), *ibid.*, Vol. 62, pp. 711-20.

[16]"Royal Order to Raja Bhupendra Singh of Bajhang," Baisakh Sudi 2, 1908 (May 1851), *ibid.*, Vol. 33, p. 173.

[17]"Reconfirmation of Narendra Bahadur Shah as Raja of Bajura," May 1887, (See n. 3 above).

[18]Yogi Naraharinath, *op. cit.*, p. 4.

[19]"Royal Order to Raja Anantapal of Bajura," Shrawan Sudi 15, 1848 (August 1791), *RRC*, Vol. 5, p. 73.

[20]"Confirmation of Dal Bahadur Shah as Raja of Achham," December 1887. (See n. 3 above).

[21]"Recognition of Prithvi Narayan Shah as Raja of Dullu," Kartik Sudi 12, 1948 (November 1891), *RRC*, Vol. 50, pp. 21-24.

[22]"Recognition of Prithvipati Khan as Raja of Garhunkot," Ashadh Badi 14, 1924 (July 1867) *ibid.*, Vol. 47, p. 124.

[23]"Recognition of Kirti Bahadur Singh as Raja of Thalahara," Poush Sudi 11, 1928 (January 1872), *ibid.*, Vol. 55, pp. 657-59.

[24]"Confirmation of Dal Bahadur Shah as Raja of Achham," December 1887, (See n. 3 above).

[25]"Recognition of Prithvi Bam Malla as Raja of Galkot," Magh Sudi 13, 1924 (February 1868), *RRC*, Vol. 64, pp. 49-67.

[26]"Recognition of Lalit Bahadur Khan as Raja of Bhirkot," Kartik Sudi 4, 1925 (October 1868), *ibid.*, Vol. 64, pp. 108-112.

[27]Satish Kumar, *Rana Polity in Nepal*, Bombay, Asia Publishing House, 1967, pp. 159-60. Jung Bahadur was given the title of *Maharaja*, whereas the chiefs of the other *rajyas* were mere *rajas*. In 1869, Dal Bikram Shah, a prince of the displaced royal family of Kaski, was granted homested lands fetching a revenue of Rs 50 a year under *birta* tenure, as well as the honorific title of *raja*. "Recognition of Dal Bikram Shah as Raja," Magh Sudi 2, 1825 (January 1869), *RRC*, Vol. 64, pp. 112-24.

[28]Regmi, *Land Tenure and Taxation*, *op.cit.*, Vol. 2, pp. 20-23.

[29]*Ibid.*, pp. 30-44.

[30]*Ibid.*, pp. 153-54.

[31]"Grant of Naya Muluk as Birta to Prime Minister Jung Bahadur and his Brothers," Bhadra Badi 4, 1925 (August 1868), in Yogi Naraharinath, *op. cit.*, pp. 177-79.

[32]'Rajkaj ko Ain," (State-affairs act), secs. 2-3.

[33]"Birta Grant to Prime Minister Jung Bahadur," Baisakh Sudi 5, 1918 (May 1861), *RRC*, Vol. 33, pp. 434-36.

[34]"Birta Grant to Prime Minister Jung Bahadur," Falgun Sudi 13, 1930 (March 1874), *ibid.*, Vol. 47, p. 230.

[35]Prime Minister Ranoddip Singh, for instance, acquired a *birta* grant in his name of lands confiscated from Major Captain Sangramsur Bisht Chhetri, a victim of the 1846 Kot Massacre, "even though the grant is in contravention of Section 3 of the State-affairs Act and the Law on Judicial Procedure of the Muluki Ain." "Bakas Birta Grant to Prime Minister Ranoddip Singh," Jestha Badi 4, 1940 (May 1883), *ibid.*, Vol. 47, pp. 673-79.

[36]These statements are based on a survey of *birta* grants made during the

period from 1846 to 1865. The practice of making *birta* grants to promote land reclamation and settlement continued in subsequent years. In 1875, for instance, Prime Minister Jung Bahadur granted waste lands amounting to one hundred *bighas* to each *Chaudhari* of the Dang-Deukhuri region in appreciation of the services rendered by them in constructing fords, tracks and irrigation channels and promoting land reclamation and settlement. (Kamal Dikshit [ed], *Battis Salko Rojanamcha [Jung Bahadurko]*. [Diary of Prime Minister Jung Bahadur's tour of the Western Tarai and of the Visit of the Prince of Wales in 1875], Lalitpur, Jagadamba Prakashan, 2023 [1966], p. 41).

[37]Ministry of Law and Justice,"Jagga Jaminko" (On land), Sec. 3, in Ministry of Law and Justice, *Shri 5 Surendra Bikram Shahdevaka Shasankalma Baneko Muluki Ain* (Legal Code enacted during the reign of King Surendra Bikram Shah Dev), Kathmandu the Ministry, 2022 (1965), pp. 19-20; "Order regarding Birta Grant to Dittha Muktinath Padhya," Baisakh Badi 6, 1926 (April 1869), *RRC*, Vol. 55, pp. 410-17.

[38]"Order regarding Irrigation Channel on Birta Lands Granted to Guru Purohit Ambar Raj Panditju," Shrawan Badi 4, 1919 (July 1862), *ibid.*, Vol. 29, pp. 482-83.

[39]Government of Nepal, "Jagga Jamin Goswarako," (On miscellaneous land matters), sec. 471, in *Ain* (Legal Code), Kathmandu, M; noranjan Press, 1927 (1870), pt. 1, p. 4. "In case any government employee grants a cash salary even though lands are available for assignment as Jagir, except on the orders of the Prime Minister, he shall be punished with a fine amounting to 2 per cent of the value of the land, and dismissed from service. The Cash salary shall be replaced by a *Jagir* land assignment."

[40]In 1872, for example, a proposal to replace the *jagir* land assignments of the Taradal Battalion in Salyan and Pyuthan by cash salaries was rejected on the ground that "salaries will have to be paid to the employees deputed to collect taxes if the lands are not assigned as Jagir, but are retained by the government." "Order regarding *Jagir* Land Assignments to Taradal Battalion," Aswin Badi 30, 1929 (September 1872), *RRC*, Vol. 44, pp. 118-20.

[41]"Regulations of the Sadar Dafdarkhana Office," sec. 17, cited in: "Order regarding Replacement of Cash Salaries by Jagir-Land Assignments," Marga Badi 30, 1935 (November 1878), *ibid.*, Vol. 64, pp. 206-13.

[42]Cf. "Order regarding Payment of Salaries to Bhairavanath Battalion," Bhadra Sudi 1, 1928 (September 1871), *ibid.*, Vol. 47, pp. 305-6.

[43]Regmi, *Land Tenure and Taxation*, *op. cit*, Vol. 3, pp. 17-18.

[44]Figures compiled on the basis of the land and revenue records of the government of Nepal for the relevant years.

[45] "Order regarding Abolition of Jagir Lands in the Tarai Region," Chaitra Sudi 2, 1909 (March 1853), *RRC*, Vol. 62, pp. 698-99.

[46]*Ibid.*

[47]*Ibid.*

[48]These levies included the *Chhalahi* and the *Jhara*. The *Chhalahi* levy was paid in the form of hides and skins by hunters, leather-workers. and members of other communities who were customarily allowed to eat the carcasses of dead cattle. *Jhara* meant the obligation to provide unpaid labour to the government. Both levies were occasionally commuted into cash payments. The practice of in-

cluding revenue from these levies in *Jagir* assignments was discontinued in 1854. "Order regarding Resumption of Revenue from Jhara and Chhalahi Levies," Aswin Badi 13, 1912 (September 1853), *RRC*, Vol. 66, pp. 11-17.

[49]"Order regarding Resumption of Revenue from Sair and Other Duties," Chaitra Badi 2, 1921 (March 1865), *ibid.*, Vol. 49, pp. 468-69.

[50]Ministry of Law and Justice, "Rukh Katnya" (On felling trees), sec. 3, in Ministry of Law and Justice, *Shri 4 Surendra . . . Muluki Ain*, p. 157.

[51]"Kathmahal Regulations for the Eastern Tarai," Marga Badi 6, 1918 (November 1861), *RRC*, Sec. 10. Vol. 10, p. 257.

4

The Agrarian Tax System

The previous chapter showed how the main financial benefit of land grants made under the *rajya*, *birta*, and *jagir* systems stemmed from the right conceded by the state to the landowning elites to appropriate a share in the peasant's produce. It would be pertinent in this context to ask what percentage of his total income the peasant was compelled to part with in fulfilment of his obligation to these elite groups, but such a quantitative anaylsis is not possible in the present state of our knowledge. We shall, therefore, confine ourselves to a broad discussion of the payments collected from the peasant, and changes in the incidence of such payments after the commencement of Rana rule.

A note on the terminology to be used for the purpose of this discussion appears necessary. The portion of his produce that a peasant pays to the government is tax, that is, a compulsory payment levied for the support of a government. Similar payments, when made by the peasant to an individual *raja*, *birta*-owner or *jagirdar* to whom the state alienated the land tilled by him, may be described as rent, that is, a specific payment made for the temporary possession or use of a house, land or other property. From the viewpoint of the peasant, however, there is no real distinction between taxes and rents, because, in either case, they are the first charge on his produce. Moreover, some peasants might be making payments to the government, whereas others in the same village might owe obligations of a similar nature to an individual landlord. In the present study, therefore, we shall describe all assessments in kind, even if they were actually collected in cash, as rents, and all assessments in money as taxes, irrespective of who actually received the payment; a *raja*, *birta*-owner, *jagirdar*, or the government.

Agricultural Rents and Taxes

Chapter 1, had discussed how agrarian systems and institutions in nineteenth-century Nepal differed according to land use (rice lands and homesteads) and location (the Tarai, the Baisi region, and the central hill region). Agricultural rents and taxes were no exception to this rule. It is impossible in the present state of our knowledge to make precise quantitative estimates. Even then, it would appear correct to generalize that during the early nineteenth century the rice-land tax absorbed approximately one-third of the produce in the Tarai and Baisi regions, and at least half in the central hill region.

In the Tarai region, the land tax was usually assessed and collected in cash. The rates were different for different crops. In Mahottari district, for example, the lowest rate of tax was twelve *annas* per *bigha* of lands under millets and lentils, whereas the highest rate of Rs 6¼ was collected on lands growing tobacco. Lands on which rice was cultivated were taxed at rates ranging from two or four rupees.[1] During the early years of the nineteenth century, paddy sold at about two maunds per rupee in the eastern Tarai region,[2] hence a tax of four rupees means an incidence of eight maunds per *bigha* against an average yield of about 25 maunds per *bigha* of rice land.[3] There were, of course, variations in the rate of tax and yields in different parts of the Tarai region, as well as both seasonal and secular changes in prices, but it would appear safe to assume that the rice-land tax absorbed roughly one-third of the produce. During the early 1840s, the figure increased to approximately 40 per cent of the produce because peasants were compelled to pay taxes on an area which was one-fourth more than the area that they actually cultivated.[4]

In the Baisi region, the rate of the rice-land tax appears to have averaged two to four *annas* per *muri*,[5] obviously according to the productivity of the land. The official procurement price of rice in Jumla was then eight *pathis* per rupee.[6] Assuming that 20 *pathis* of paddy yielded 10 pathis of rice, this would mean a price of 16 *pathis* of paddy per rupee. The tax, therefore, amounted to maximum of four *pathis* per *muri* on rice-land, or 16 *pathis* per *ropani*. Assuming a maximum yield of about 50 *pathis* of paddy per *ropani*, this would mean, as in the Tarai region, an incidence of approximately one-third of the produce.

In the central hill region, on the other hand, the rice-land rent was assessed and collected in the form of paddy and amounted to at least half of the produce under what was known as the *adhiya* system. Under this system, the actual produce of a field was divided equally between the landlord and the peasant. On those categories of rice lands on which rents were payable not to *jagirdars* or other individuals, but directly to the government, rent was usually assessed under the *kut* system in the form of a specific quantity of paddy,[7] or a sum of money,[8] irrespective of the actual yield. Available evidence suggests that *kut* rents during the late eighteenth and early nineteenth centuries were at least as high as *adhiya* rents, and were fixed at specific rates mainly to facilitate collection without sharing the actual produce on the threshing-ground itself.

In addition to rents under the *adhiya* or *kut* system, peasants in the central hill region were under obligation to pay at least three cash levies on the rice lands they tilled. These were the *chardamtheki*,[9] payable on the renewal of the peasant's tenure on the land every year, a tax on winter crops, and the *ghiukhane*,[10] which was meant to finance the supply of *ghee* for the landlord's household. All these levies were paid in cash. Another levy, which was collected on rice lands in Kathmandu Valley only, was meant to finance the salaries of watchmen employed to protect crops from stray cattle.[11] There was also one levy to finance the maintenance of state-owned irrigation channels which irrigated rice-lands in Kathmandu Valley. However, these two payments were more in the nature of fees for services rendered than taxes without a *quid pro quo*.

Extension of the Kut System

After the second decade of the nineteenth century, the *adhiya* system was gradually replaced by the *kut* system throughout the central hill region, thereby setting off a trend toward the maximization of the rice-land rent. The *adhiya* system contained several shortcomings from the viewpoint of *jagirdars*. Because their income consisted of a share of the produce, it was liable to fluctuate from year to year. If crops were good, they collected an adequate income. If, however, crops failed, they were in no position to make both ends meet throughout the year. Rents, no doubt, were the first charge on the produce, but a charge that varied with the yield. There was also little the *jagirdar* could do if peasants cheated in the sharing of crops, or cultivated the land negligently, or even did not cultivate it at all.[13]

In contradistinction, *kut* rents were a specific charge on the produce which normally could not be reduced to suit the harvest, and so yielded a more stable income.[14] Because of this advantage, the *kut* system was adopted in 1812 as the basis of rent assessment on rice-lands under *jagir* tenure throughout the central hill region. This measure had a twofold objective: to avoid the uncertainty in the quantity of rents paid under the *adhiya* system because of the cultivator's dishonesty or negligence, and to fix rents at a standard level determined by the potential yield of the land,[15] rather than by its actual yield from year to year as under the *adhiya* system.

The *kut* system also permitted the peasant to meet his fiscal obligations in money, if he so desired.[16] Hamilton, obviously referring to the introduction of the *kut* system in Kathmandu Valley in 1812, pointed out that each farm was assessed at a certain quantity of grain, which the farmer might either pay in kind, or in money, at the market price.[17] The commutation of *kut* rents was, in fact, so convenient for *jagirdars* that often preference in the allotment of rice lands was given to peasants who were willing to make payments in cash.[18] The practice seems to have been widely followed in the course of the 1836-37 revenue settlements. The rates of commutation varied in different areas, apparently on the basis of the distance from Kathmandu.[19] They were usually fixed on a long-term basis; hence any rise in the price of agricultural produce reduced the real burden of the tax on the peasant. This possibly explains why at times in-kind *kut* rents were commuted into cash on the peasant's own initiative.[20]

From the viewpoint of agricultural production, the transition from proportional rents under the *adhiya* system to fixed rents under the *kut* system may have had a favourable effect on agricultural production. Proportional rents lessen the incentive of the farmer to adopt improvements, inasmuch as for a given improvement to be worth while at the margin to the farmer it must yield twice as much if the rent were a fixed amount.[21] In contrast, the secure variants that a fixed-rent system such as *kut* provides tend to improve both resources allocation and productivity.[22] In other words, an *adhiya* farmer must share with his landlord any increase in production that he is able to achieve through more intensive cultivation, irrigation, etc. On the other hand, because his fiscal obligations to the landlord have been determined in advance, a farmer working under the *kut* system can fully appropriate the benefits of increased production.

The introduction of the *kut* system on rice-lands in the districts of the central hill region during 1812 seems to have generally had this effect. By the mid-1820s, average yields had apparently increased to a level which made the fixed rents payable under the *kut* system actually less than half of the crop. Consequently, there was pressure from *jagirdars* to raise *kut* rents so as to keep pace with the increased yields. In 1828, therefore, the government sanctioned a general increase in the level of *kut* rents in all parts of the central hill regions.[23] Tenants who refused to pay the enhanced rents were liable to eviction.

A rough survey of *kut* rent assessments made in different parts of the central hill region during the 1836-37 settlements show that the rates ranged between 20 and 30 *pathis* of paddy per *ropani*. The normal yield of paddy in that region was then 80 *pathis* per *ropani*.[24] The yield was naturally lower on lands with inferior soils or poor irrigation facilities, and probably amounted to no more than 40 *pathis* per *ropani*.[25] This would mean that *kut* rents absorbed between one-half and two-thirds of the total paddy crop.

The main aspects of the *kut* system that need to be emphasized in the context of the present study are the higher level of *kut* rents compared with *adhiya* rents and their temporary nature. Inasmuch as *jagirdars* had full authority to give away their lands for cultivation to any peasant who offered to pay the highest amount of rent in any year, the level of *kut* rents was not only high but also uncertain.

Homestead-Tax System

Chapter 1 had shown that homestead lands constituted a separate category for purposes of taxation in the hill region only. Taxes were assessed according to the estimated size of homesteads in terms of the number of ox-teams needed for ploughing.[26]

There were at least two taxes on homesteads in the hill region;

the *saunefagu*, which was levied on each roof, and the *serma*, which was based on a rough estimation of the size of the homestead.[27] Both these taxes were in cash. The rate of the *serma* tax ranged between eight *annas* to one rupee and the *saunefagu* averaged one *anna* on each homestead.[28]

Additional Levies

In addition to the *serma* and the *saunefagu*, the peasant in nineteenth-century Nepal was under obligation to pay a multitude of levies. Many of these levies were in the nature of a poll-tax, or were based on caste, ethnic origin or occupation. The number and nature of these levies were different in different parts of the country, and, indeed, even in different villages of the same district. An attempt has been made in earlier studies on the subject to enumerate these levies and discuss their nature and origin,[29] but the information is still too meager to make it possible to determine the total incidence on each individual homestead. For the purposes of the present study, therefore, it may be sufficient to stress that such taxes and levies cumulatively imposed a real burden on the peasant's homestead income.

Jagirdars and other landlords were usually entitled "to collect customary gifts and presents from time to time" from the inhabitants of villages and areas under their control.[30] The nature of these gifts and presents was nowhere defined, hence the right meant nothing less than a blank check on the peasant. The result was that peasants were under obligation to provide to the landlord, in addition to the *serma* and *saunefagu* taxes on their homestead, such miscellaneous items as fruits, vegetables, *ghee*, oil, eggs, fish, chicken, firewood, and bamboo poles, and sometimes even such manufactured articles as caps and shoes.[31] Often the combined value of these additional payments exceeded the amount of the *serma* and the *saunefagu*. Flagrant infractions of custom in the collection of such payments were often brought to the notice of the government and remedied,[32] but a villager could hardly approach Kathmandu each time the *jagirdar* or his agent demanded "gifts and presents."

Although homestead taxation of the kind prevalent in the hills was usually unknown in the Tarai region, there is evidence that special levies were collected in consideration of the peasant's use of such common facilities as forests and sources of water.[33] Peasants were also under obligation to finance religious ceremonies at the village shrine, as well as revenue settlement operations, the cost of stationery and furniture used in local administrative offices[34] and the maintenance of the state elephants.[35]

The Asmani Tax

Nineteenth-century Nepali revenue documents use the term *asmani* to denote the income accruing from judicial fines and penalties. The inclusion of such income in a discussion of agrarian taxation perhaps requires an explanation. Chapter 3 had referred to the rights of *rajas*, *birta*-owners and *jagirdars* to administer justice in the areas under their jurisdiction. That right was important to them for two reasons: it enabled them to maintain effective control over their subjects, and more important, it fetched them income in the form of *asmani* and manpower through the enslavement of criminals. Indeed, such income formed an important component of the financial benefits that a *raja*, *birta*-owner, or *jagirdar* obtained from the lands and villages under his jurisdiction. The right to dispense justice was, therefore, valued primarily because it was a source of income, and justice was usually dispensed in a manner that maximized these benefits.

Modern jurisprudence regards the imposition of fines as a form of punishment for infractions of the law. In nineteenth-century Nepal, however, every individual was regarded as a potential law-breaker and fined in advance. In other words, peasants were compelled to bear the burden of *asmani* payments whether or not they actually committed a wrong. In some parts of the region, for instance, *asmani* payments were assessed in advance at a specified percentage of the *serma* tax on the homestead. The percentage ranged between 50[36] and 200.[37] Eventually, *asmani* assumed the form of a separate tax on the homestead, which was collected in addition to any fines that might be imposed on the peasant for actual wrong-doing.[38] The landowning elite's right to dispense justice, consequently, degenerated to one more avenue for squeezing the peasant.

Extra-Legal Collections

No account of agricultural taxation in nineteenth-century Nepal can be complete without a reference to the extortionate collections that local officials and visiting dignitaries usually exacted from the peasantry. The chronic and ubiquitous character of these exactions made them an inseparable part of the agricultural tax system and raised taxation to a level far above what the formal tax system alone would indicate. For example, in districts where military headquarters were situated, such as Ilam and Doti, the procurement of provisions to feed the troops imposed an additional burden on the peasantry. Local military authorities paid for food supplies at prices lower than those current in the village, or even collected them forcibly without any payment.[39] Moreover, practically all over the country, the burden of feeding local functionaries and visiting dignitaries fell on the peasant.[40] Indeed, there is evidence that visiting officers and dignitaries frequently extorted supplies and provisions from the peasantry without any payment. Similarly, in the Tarai districts, "besides the formal or established cess, the *jamindar* or cultivator is obliged to pay occasionally, other irregular and arbitrary taxes in the form of fines, doucers, and the like.[41]

Unfair Collection Procedures

Extra-legal exactions were not the sole factor that enhanced the burden of taxation on the peasant beyond the level set by formal assessment rates. The government often also followed collection procedures which were patently unfair to the peasant. For instance, peasants were often required to provide free porterage services for the transportation of goods which they were under obligation to supply against the homestead tax revenue due from them. The inhabitants of several villages in Pyuthan district were under obligation to pay a part of the revenue assessed on their lands and homesteads in the form of saltpeter, sulphur, and other materials required for the manufacture of gunpowder. To procure these commodities, they had to travel as far as Nepalgunj in the far-western Tarai, and even to the adjoining areas of India,[42] but the conversion rate took no account of that additional obligation.

A similar squeeze on the peasantry resulted from the practice of procuring manufactured commodities with the homestead-tax revenue due from a village. The terms of procurement were usually unfavourable for the peasant, as otherwise there would have been no incentive to the government to adopt such a procurement policy, but the practice in effect increased the tax-burden in real terms. For instance, Khinchet and Buntang villages in Nuwakot district were under obligation to supply their entire production of paper to the government at a price which was about 60 per cent less than the current market price.[43] The cost of the supplies was adjusted against the homestead taxes due from them[44] so that the craftsmen actually paid their taxes in the form of a manufactured commodity at an unfavourable price. They were, no doubt, provided with credit,[45] and allowed to procure raw materials free of cost,[46] but these concessions do not seem to have been a sufficient compensation for the unfavourable price differential. Payment of money taxes in commodites accordingly deprived the peasantry of the opportunity of getting higher prices in the market.

Impact of High Level of Taxation

By the middle of the nineteenth century, haphazard tax enhancements and a multitude of *ad hoc* and arbitrary levies on rice lands and homesteads had cumulatively imposed a heavy burden on the tax-paying capacity of peasants, particularly in the central hill region. Arbitrary enhancements of *kut* rents and the eviction of peasants who were not in a position to accept such enhancements became a common-place occurrence. Moreover, rice lands often remained uncultivated because the increased *kut* rents stipulated by land-hungry peasants were not always justified by actual yields.[47] The situation that prevailed in the Tarai region was possibly less bleak because low population density fostered competition among landlords for prospective tenants, rather than among the peasantry for allotments of land. Even then, there is evidence that many peasants abandoned their holdings in an attempt to avoid paying the 25 per cent surcharge imposed during the early 1840s.[48]

The nature of the problems that the Rana government faced in the field of agricultural taxation was thus different in different regions of the Kingdom. The basic problem related to the temporary and hence uncertain nature of rent assessments under the *kut* system in the central hill region. In that region, therefore, the Rana government made an attempt to relate the assessments more closely with productivity, and fix them on a long-term basis. In the Tarai region, the main thrust of policy was toward abolishing arbitrary levies and simplifying the land-tax assessment system.

Reforms in the Kut System

In order to end the haphazard and uncertain nature of rent assessments under the *kut* system in the central hill region, the Rana government enforced legislation in 1854 abolishing the right of landlords to raise *kut* rents. A *jagirdar* was no longer permitted to evict his tenant on the ground that another peasant had offered a higher rent for his *jagir* lands. Thanks to that measure, the rights of a peasant remained secure so long as he paid the stipulated *kut* rents.[49] With the objective of breaking the *jagirdar's* control over the peasant more effectively, the Rana government utilized the *tirja* system on a more extensive basis. *Tirjas*, as noted in Chapter 2, were certificates issued by the central authorities which authorized a *jagirdar* to collect rents on the lands assigned to his as *jagir*. These certificates specified the name of the peasant, the area and location of the holding, and the form and amount of the rent. Thanks to the *tirja* system, the obligations of a peasant who cultivated rice-lands under *jagir* tenure were clearly specified; in no circumstances was the *jagirdar* permitted to collect higher or additional payments on rice lands. Indeed, a total ban was imposed on rent enhancements during the interval between two revenue settlements.

Although the 1854 legislation prevented arbitrary rent enhancements on *jagir* lands, it was not retroactive. Problems created by such enhancements in the past, therefore, continued. During the period from 1854 to 1868, settlements under the *kut* system were revised throughout the central hill region, and the level of rents was reduced where necessary. Inasmuch as the 1854-68 settlement were not revised in most districts of the central hills region during the remaining years of the nineteenth century, the level of *kut* rents remained more or less unchanged.

During the 1890s, the Rana government completed the transition from the *adhiya* system to the *kut* system of rent assessment in the central hill region. Existing *adhiya* rents were then converted into *kut* at the level of actual collection, or of the *kut* rents being collected on adjoining holdings, whichever was higher.[50] Thanks to that policy, the highest rate prevailing in any area determined the general level of rents in that area. The *adhiya* system was, thereafter, retained only in exceptional circumstances where it appeared that any increase in rents would remove lands from cultivation.[51]

Long-term settlements were not the sole factor that provided a measure of rigidity to *kut* rents in the central hill region during the Rana period. Efforts were also made to insulate *kut* rents to some extent from the uncertain impact of fluctuations in yields caused by the failure of crops. In the central hill region, for instance, the following formula was introduced in 1854 for remissions of *kut* rents in the event of crop failures:[52]

> The amount of *kut* rent shall be doubled to calculate the gross yield. If the loss amounts only to one-fourth of that yield, no remission shall be allowed. If the percentage of loss is higher, the actual produce shall be divided on *adhiya* basis.

A minor decline in production consequently had no effect on the fiscal obligations of the peasant. For example, if the *kut* rent on a plot of rice land amounted to 30 *pathis* of paddy, the gross yield was calculated at 60 *pathis*. If the actual yield amounted to no more than three-fourths of that quantity, that is, 45 *pathis*, the peasant was not entitled to any remission, but would still be required to pay 30 *pathis*. Only if the damage exceeded one-fourth of the normal production was the peasant entitled to demand that the crop be divided under the *adhiya* system.

Simplification of the Tax-Assessment System in the Tarai

As noted previously, Rana policy aimed at simplifying the tax-assessment system in the Tarai. The main objective of that policy was to facilitate the task of tax collection. During the period from 1849 to 1857, a number of additional levies customarily collected in the districts of the eastern Tarai region were abolished. These levies included those on common lands and sources of water,[53] and the 25 per cent surcharge on the cultivated area contained in peasants' holdings. Several other levies were consolidated into a single payment with the basic land tax.[54]

Efforts were also made to establish a more accurate correlation between tax assessment rates and actual yields, as in the central hill region. For instance, lower rates were prescribed in the northernmost strip adjoining the foothills of the Siwalik range than in the southern areas adjoining the Nepal-India border.[55] Moreover, because of the need to attract settlers to colonize waste lands in the Tarai region, rates were usually kept at a low level, and seldom increased without the consent of the peasantry.[56] At the same time, the Rana government seldom accepted any reduction in the level of land tax-revenue. Once a field was measured and taxed, no remission was given on any account.[57] Peasants were not permitted to replace high-tax crops by low-tax ones, but taxes increased if low-tax crops were replaced by high-tax ones.[58]

If we ignore the abolition of the short-lived surcharge on the cultivated area in the eastern Tarai region, and also the minor adjustments in tax assessments that were made in individual cases in the central hill region, it would appear correct to generalize that the incidence of taxation in all parts of the country remained more or less the same under Rana rule as before. The chief achievement of Rana policy in the field of land-taxation was the introduction of long-term tax-assessment rates in all parts of the country, but the Ranas also followed collection procedures which lent a measure of rigidity to the land-tax system. Consequently, they were able to squeeze a higher percentage of the peasant's food output in bad years than in good ones. It is also significant that the Ranas made little effort to reform the homestead-tax system in the hill region. On the contrary, they improved both the machinery and the procedure for the collection of homestead taxes, with the consequence that the incidence of such taxation, at least in monetary terms, increased to an unprecedented level. We shall discuss the tax-collection system, and the role of the village elites under that system, in Chapter 5.

Agricultural Rents on Birta Lands

The foregoing sections presented a skeletal outline of the land-tax reform measures undertaken in different regions of the country after the emergence of Rana rule. The outline covered only *jagir* and other taxable lands on which documentary evidence is available. Many peasants also cultivated *birta* lands in the capacity of tenants and share-croppers, but hardly any effort was made before 1854 to regulate landlord-tenant relations on *birta* lands.[59]

It has been noted earlier that the 1854 legal code abolished the traditional right of *jagirdars* to increase rents on their *jagir* lands and evict tenants who did not agree to pay rents at the enhanced rates. In contradistinction, the code reconfirmed the traditional right of *birta*-owners to evict their tenants, and resume their lands and homesteads for personal cultivation and residence.[60] *Birta-*

owners were also allowed to take advantage of competition among prospective tenants to maximize their income. A tenant cultivating a plot of *birta* land could be evicted whenever another person offered to pay a higher rent which he himself was unable to pay, except when the peasant had reclaimed the land himself.[61] Accordingly, there was no limit to the amount of rent *birta*-owners could collect from their tenants. The law recognized the validity of any agreement that might be concluded between *birta*-owners and their tenants regarding the rate of rents and other conditions of tenancy,[62] but did not take into account the peasant's need for a subsistence holding and his low bargaining power.

The power that the *birta*-owner wielded over his tenants was probably much less than what the provisions of the 1854 legal code suggest. Hardly any *birta*-owner possibly risked violating the customary law which stipulated that no peasant should be deprived of his holding before he had harvested the crops that he had sown. A more effective check on the *birta*-owner's power over his tenants possibly stemmed from the competition that the government itself offered. Peasants were subject to fiscal and other constraints if they abandoned their *raikar* land allotments and shifted to *birta* lands. *Birta*-owners were forbidden to attract peasants from *raikar* lands. On the other hand, individuals who started *raikar* land-reclamation projects, or obtained contracts for the collection of taxes on *raikar* lands, were encouraged to attract peasants from *birta* lands by every possible means. A *birta*-owner who tried to impose conditions of tenancy harsher than those prevalent on *raikar* lands in the area was, therefore, in danger of losing his tenants altogether. Consequently, rents and taxes on rice lands and homesteads under *birta* tenure were normally fixed on the basis of rates prevalent on adjoining *raikar* holdings,[63] and hence were approximately equal.[64]

We may conclude that the main characteristic of the agrarian tax system in nineteenth-century Nepal was the multifarious and burdensome nature of the payments in money or in kind that the peasant had to make on an obligatory basis on both his rice lands and homesteads to *rajas*, *birta*-owners, *jagirdars*, or the government. All these payments had to be treated as prior charges: they could seldom be adjusted to suit the actual yield of the land or the personal circumstances of the peasant. It was with the residual income, left after meeting all these prior charges, that the peasant could meet his subsistence needs. The agricultural tax system was devised mainly with the objective of ensuring a steady income for the landowning elites, or revenue for the government. The system left the peasants at a minimum level of subsistence, and took little account of the need to reinvest at least a portion of the agricultural surplus to raise the standard of living of the peasant, effect improvements in the land, and thereby raise the level of agricultural productivity.

NOTES

[1]"Confirmation of 1793 Tax-Assessment Rates in Mahottari," Kartik Sudi 10, 1866 (November 1809), *RRC*, Vol. 40, pp. 111-18.

[2]"In autumn 1809, for instance, there being at Nathpur a great demand for grain in order to send it to Patna, the merchants made large advances to the farmers of Suban Saptari, where there was a good crop, and agreed to give a rupee for two *mans* of rice in the husk." Francis Buchanan (Hamilton), *An Account of the District of Purnea in 1809-10*, Patna, Bihar and Orissa Research Society, 1928, p. 576.

[3]Government of Nepal, "Industrial Survey Reports for Bara, Parsa, Rautahat, Sarlahi, and Mahottari," Kathmandu, Department of Industrial and Commercial Intelligence, 2005 (1949), p. 16.

[4]"Order to Subedar Badrinath Chhatkuli Regarding Complaints of People of Rautahat District," Marga Sudi 2, 1902 (November 1845). *RRC*, Vol. 7, p. 443.

[5]Cf., "Thek-Theti Arrangements for Gam-Dara in Jumla District," Chaitra Sudi 7, 1900 (March 1844), in Yogi Naraharinath (ed.), *Itihas Prakash* (Light on History), Kathmandu, Itihas Prakash Mandal, 2012-13 (1955-56), Bk 2, Vol. 2, pp. 273-84.

[6]"Order Regarding Revenue Collection in Jumla," Ashadh Badi 5, 1891 (June 1834), *RRC*, Vol. 40, pp. 491-97.

[7]Cf., "Kut Rent Assessment on Rice-Lands of Bale Padhya in Nuwakot," Marga Badi 30, 1853 (November 1796), *ibid.*, Vol. 5, p. 351.

[8]Cf. "Kut Rent Assessment on Rice-Lands of Sarbajit Karki in Nuwakot," Kartik Sudi 5, 1851 (November 1294), *ibid.*, Vol. 5, pp. 297-98.

[9]Mahesh C. Regmi, *A Study in Nepali Economic History*, 1768-1846, New Delhi, Manjusri Publishing House, 1971, p. 85.

[10]*Ibid.*, p. 85.

[11]"Sahanapiwal Levy Regulations," Bhadra Badi 14, 1900 (August 1843), *RRC*, Vol. 33, pp. 504-9.

[12]Cf., "Royal Order to Dhalwas in Kirtipur," Aswin Sudi 15, 1886 (October 1839), in Chittaranjan Nepali, *Janaral Bhimsen Thapa ra Tatkalin Nepal* (General Bhimsen Thapa and Contemporary Nepal), Kathmandu, Nepal Sanskirtik Sangh, 2013 (1956), pp. 208-9,

[13]"Orders Regarding Harvesting of Crops on Jagir Lands," Separate orders for Kaski, Parbat, Lamjung, Bhirkot, etc. Shrawan Badi 1, 1864 (July 1807), *RRC*, Vol. 20, pp. 428-29; "Order Regarding Kut Rent Assessment in Bungmati, Lalitpur," Aswin Badi 6, 1866 (September 1809), *ibid.*, Vol. 40, p. 103.

[14]Cf., "Complaint Regarding Kut Rent Assessment on Hulak Lands in Palanchok," Ashadh Badi 9, 1921 (June 1864), *ibid.*, Vol. 55, pp. 98-99.

[15]Cf. "Order Regarding Kut Rent Assessment on Jagir Lands of Jwaladal Company," Ashadh Badi 13, 1882 (June 1825), *ibid*, Vol. 34, p. 14.

[16]"Order Regarding Kut Rent Payment on Jagir Lands of Ranadal Company in Musikot," Baisakh Badi 7, 1885 (April 1828), *ibid.*, Vol. 43, pp. 85-86.

[17]Francis Hamilton, *An Account of the Kingdom of Nepal* (reprint of 1819 ed.) New Delhi, Manjusri Publishing House, 1971, p. 211.

[18]"Order Regarding Rice Land Allotments in the Chepe-Marsyangdi-Bheri Region," Poush Sudi 6, 1885 (January 1829), *RRC*, Vol. 43, pp. 161-62.

[19]Cf., "Order Regarding Commutation of Kut Rents in Dhading," Jestha Sudi 7, 1896 (June 1839), *ibid.*, Vol. 38, p. 47.

[20]*Ibid.*

[21]W. Arthur Lewis, *The Theory of Economic Growth*, London, George Allen & Unwin Ltd. 1963, p. 123.

[22]Clive Bell, "Ideology and Economic Interests in Indian Land Reform," in David Lehmann (ed.), *Agrarian Reform & Agrarian Reformism*, London, Faber and Faber, 1974, pp. 198-99.

[23]Regmi, *Nepali Economic History, op. cit.* p. 180.

[24]William Kirkpatrick, *An Account of the Kingdom of Nepaul* (reprint of 1811 ed.) New Delhi, Manjusri Publishing House, 1969, p. 95; Hamilton, *An Account of the Kingdom of Nepal, op. cit.*, pp. 216 and 225.

[25]Cf., Government of Nepal, "Industrial Survey Reports for West Nos. 1 and 3," Kathmandu, Department of Industrial and Commercial Intelligence, 2003 (1946), p. 18.

[26]Mahesh C. Regmi, *Land Tenure and Taxation in Nepal*, Berkeley, University of California Press, 1963-68, Vol. 1, p. 64.

[27]*Ibid.*, pp. 43-48 and 73-74.

[28]*Ibid.*

[29]*Ibid.*, pp. 43-48 and 74-75.

[30]"Order Regarding Privileges of Jagirdars," Falgun Badi 9, 1893 (February 1837), *RRC*, Vol. 40, p. 633.

[31]Regmi, *Land Tenure and Taxation, op. cit.*, Vol. 1, pp. 197-99.

[32]"Complaint Against Dware Manbir of the Chautara Iron Mines," Falgun Sudi 15, 1949 (March 1893), *RRC*, Vol. 57, pp. 329-41.

[33]"Confirmation of 1793 Tax Assessment Rates in Mahottari," 1809, (See n. 1 above). See also, "Some Errors in Francis Hamilton's 'An Account of the Kingdom of Nepal," *Regmi Research Series*, Year 6, no. 3, March 1, 1974, pp. 41-45.

[34]*Ibid.*

[35]"Order Regarding Dispersal of Hattisar Offices in the Tarai Region," Aswin Sudi 6, 1919 (September 1862), *RRC*, Vol. 33, pp. 546-47.

[36]Cf., "Thek-Thiti Grant for Revenue Collection in Sumle village, Kaski

District," Ashadh Badi 11, 1894 (June 1837), *ibid.*, Vol. 34, pp. 328-30.

[37]Cf., "Thek Grant for Revenue Collection on Jagir Lands of Sri Gorakh Bux Company in Udayapur," Chaitra Badi 8, 1912 (March 1856), *ibid.*, Vol. 58, pp. 14-23.

[38]Regmi, *Land Tenure and Taxation, op. cit.*, Vol. 1, pp. 45-46; "Miscellaneous Regulations for Jumla District," Falgun Badi 30, 1915, *RRC*, Vol. 29, pp. 272-79.

[39]"Thek-Thiti Arrangements for Dullu-Dailekh," Marga Badi 5, 1938 (November 1881), *ibid*, Vol. 58, pp. 129-40; "Procurement of Supplies for Troops Stationed in Doti," Bhadra Sudi 12, 1949 (September 1892), *ibid.*, Vol. 58, pp. 755-63, and Magh Sudi 10, 1949 (February 1893), *ibid.*, Vol. 50, pp. 779-96.

[40]Cf., "Prohibition to Exact Provisions from Inhabitants of Panchsaya Khola," Kartik Badi 13, 1912 (October 1855,) *ibid.*, Vol. 56, pp. 512-14. There is evidence that such extortions were commonly practised until recent times. David Snellgrove, while travelling through Kahinikanda village in the Bheri region in 1956, wrote, "Our efforts to procure eggs and potatoes met with little success, for the people took us to be officials and feared that we might refuse to pay for what they gave us." Similarly, at Tibrikot, "(The headman) . . . offered to help us in any way possible, so long as we paid for what we received. We have heard this condition stated so often, that we can only conclude that many who travel with government sanction seek to press the villagers into supplying their needs free of charge." David Snellgrove, *Himalayan Pilgrimage*, Oxford, Bruno Cassirer, 1961, pp. 17 and 26.

[41]Kirkpatrick, *An Account of the Kingdom of Nepaul, op. cit.*, p. 41.

[42]"Procurement of Sulphur and Saltpeter for Gunpowder Factory in Pyuthan," Jestha Sudi 9, 1895 (June 1838), *RRC*, Vol. 34, pp. 447-52, 459-68, and Jestha Badi 11, 1921 (May 1864), *ibid.*, Vol. 49, pp. 205-17.

[43]"Order Regarding Supply of Paper from Buntang," Baisakh Sudi 2, 1954 (April 1897), *ibid.*, Vol. 61, pp. 659-86; "Order Regarding Supply of Paper from Mahabharat-Gorkha Region," Bhadra Sudi 10, 1958 (September 1901), *ibid.*, Vol. 52, pp. 9-12.

[44]"Order Regarding Supply of Paper from Buntang," Shrawan Sudi 10, 1927 (August 1870), *ibid.*, Vol. 64, p. 223.

[45]"Contract for Manufacture of Paper in Buntang," Bhadra Sudi 11, 1949 (September 1892), *ibid.*, Vol. 58, pp. 780-92.

[46]"Order Regarding Supply of Paper from Mahabharat-Gorkha Region," 1901. (See n. 43 above).

[47]Cf., "Order Regarding Complaint of Shaktiballabh Lohani of Chiti, Lamjung District," Magh Badi 12, 1921 (January 1865), *RRC*, Vol. 21, pp. 204-5; "Reallotment of Rice-Lands in Batulechaur, Kaski District," Magh Badi 11, 1894 (January 1838), *ibid.*, Vol. 27, p. 500; "Order Regarding Unauthorized Kut Rent Enhancements on Jagir Lands of Gorakh Bux Company in Udayapur," Baisakh Sudi 3, 1907 (April 1850), *ibid.*, Vol. 64, pp. 673-75.

[48]"Order to Subedar Badrinath Chhatkuli Regarding Complaints of People of Rautahat District," November 1845. (See n. 4 above).

[49]Ministry of Law and Justice, "Mohi Talsingko" (On Tenants and Landlords), Sec. 3, in *Shri 5 Surendra Bikram Shahdevaka Shasankalama Baneko*

Muluki Ain (Legal code enacted during the reign of King Surendra Bikram Shah Dev), Kathmandu, the Ministry, 2022 (1965), p. 38.

[50]"Order Regarding Conversion of Adhiya Rents into Kut," Baisakh Badi 13, 1954 (April 1897), *RRC*, Vol. 61, pp. 583-93; "Regulations in the Name of the Sadar Dafdarkhana," Magh 25, 1961 (February 7, 1905), *ibid.*, Vol. 16, pp. 532-34. This order reconfirmed arrangements prescribed in A.D. 1894.

[51]*Ibid.*

[52]Ministry of Law and Justice, "Bali Na Tirnya Mohiko" (on default in the payment of rents), sec. 7, in *Shri 5 Surendra Bikram . . . Muluki Ain*, p. 48.

[53]"List of Abolished Taxes and Levies in the Tarai Region," 1906 (1849), *RRC*, Vol. 37, pp. 3-4; "Order Regarding Abolition of Miscellaneous Taxes and Levies in the Tarai Region," Baisakh Badi 3, 1907 (April 1850), *Ibid.*, Vol. 64, pp. 509-23

[51]"Settlement Regulations for the Eastern Tarai Districts," Marga Badi 6, 1918 (November 1861), *ibid.*, Sec. 8, Vol. 10, pp. 155-56.

[55]*Ibid.*, Sec. 11, p. 157.

[56]"Order Regarding Land-Tax Assessments in Rautahat District," Baisakh Sudi 4, 1942 (April 1885), *ibid.*, Vol. 56, pp. 17-38.

[57]"Settlement Regulations for the Eastern Tarai Districts," 1861, *ibid.*, Sec. 5, pp. 154-55. (See n. 54 above).

[58]*Ibid.* Sec. 17, pp. 160-61.

[59]Regmi, *Nepali Economic History, op. cit.*, p. 91.

[60]"On Default in the Payment of Rents," Sec. 1, p. 38. (See n. 52 above).

[61]*Ibid.*, Sec. 13, p. 40.

[62]*Ibid.*, Sec. 2, p. 38.

[63]"Order Regarding Settlement of Tenants on Birta Lands in Udayapur," Jestha Sudi 3, 1942 (May 1885), *RRC*, Vol. 52, pp. 335-52.

[64]"Order Regarding Satta Birta Grants in Kathmandu," Kartik Badi 3, 1924 (October 1867), *ibid.*, Vol. 64, pp. 159-77.

5

The Village Elite

Rajas, *birta*-owners and *jagirdars*, who appropriated the rents and taxes paid by peasants, seldom lived in the villages where their lands were situated. The government, which too claimed a share in the peasant's crop on lands that had not been alienated under the *rajya*, *birta* and *jagir* systems, had no tax-collection offices in the village during the nineteenth century. Consequently, not only *rajas*, *birta*-owners, and *jagirdars*, but even the government, needed the services of village headmen or other functionaries to collect agricultural rents and taxes on their behalf. These headmen or functionaries thus played an intermediary role between the landowning elites or the government and the peasant. They were seldom paid any formal emoluments for their services; rather, they were given a special status and privileges vis-a-vis the peasantry that made it possible for them to partake of a share in the agricultural surplus in lieu of emoluments. We may therefore, designate them as the village elites, or local representatives of the landowning elites of the aristocracy and the bureaucracy. The purpose of this chapter is to show that the peasants of nineteenth-century Nepal shared their produce not only with *rajas*, *birta*-owners, *jagirdars*, or the government, but also with the functionaries employed to collect that share.

Tax-Collection Functionaries

Chapter 4 had discussed the different tax-assessment systems followed on rice-lands and homesteads in the hill regions during the nineteenth century. In general, collection of taxes presented more problems on homesteads than on rice-lands. Taxes on rice-lands were assessed on the basis of the area; hence revenue from such taxes could be estimated in advance. On the other hand, the multiplicity of homestead taxes and levies, and the fluctuating number of homesteads in a village, made it difficult to estimate homestead-tax revenue in advance. Nor was this all. *Rajas, birta-*owners, and *jagirdars* exercised judicial authority over the inhabitants of villages owned by them. The exercise of that authority fetched income in the form of fines and penalties, which was an important component of the homestead-tax system. Judicial fines and penalties, although a regular source of income, naturally could not yield a definite amount each year.

Difficulties of collection were added to the indeterminate nature of homestead-tax revenue. Because the amount of revenue depended on the number of homesteads, close and regular inspection, and regular changes in the tax-assessment register, were necessary in order to make adjustments for the changing number of homesteads. Moreover, homesteads in the hill regions were usually located in widely-separated, sparsely-populated and inaccessible hamlets. The amount collected from taxes was, therefore, often less than the cost of maintaining a regular administrative agency for that purpose.

Because of these differences in systems of taxation on rice-lands and homesteads, different functionaries were employed for the collection of revenue from these two sources. Traditionally, there existed in all parts of the Kingdom a multi-tiered hierarchy of local functionaries to allot lands, maintain records of lands and tax assessments, and collect taxes. A detailed description of this hierarchy has been given elsewhere;[1] in the present context, it may be sufficient to deal with the main functionaries that existed at the village level in the hill region and the Tarai.

In the hill region, village-level tax-collection functionaries during the nineteenth century consisted of *mukhiyas* and *jimmawals.* The *mukhiya* was the village headman, a link between the central government and the local community. He was mainly responsible for the collection of homestead taxes and levies, but his jurisdiction covered almost all aspects of village life. The *jimmawal* seems to have been a less important functionary, inasmuch as his jurisdiction extended to rice-lands only. His authority was further circumscribed by the fact that *jagirdars* usually depended for the collection of rents on their rice-land assignments on brokers called *dhokres*. As already noted in Chapter 3, *jagirdars* were allowed to collect such rents only on the basis of negotiable certificates called *tirjas*. *Dhokres* purchased *tirjas* from *jagirdars* and collected rents from the peasant on the basis of that authority. Their profit consisted of the difference between the actual collection and the price at which they purchased *tirjas* from *jagirdars*.

In most parts of the Tarai region, where the distinction between rice-lands and homesteads had little significance from the viewpoint of taxation, revenue collection was the responsibility of a functionary known as the *chaudhari*. The term was customarily used to denote a big landowner, and obviously prominent local landowners were appointed to that post. The *parganna*, a subdivision of a district consisting of a varying number of settlements or villages called *moujas*, normally comprised the jurisdiction of a *chaudhari*.[2]

Tax-Collection Systems

During the early nineteenth century, four different systems were used for the collection of agrarian taxes in different parts of the country: *amanat*, *ijara*, *thekbandi*, and *thekthiti*. Three of these systems, *amanat*, *ijara*, and *thekbandi*, were used for the collection of homestead taxes in the villages of the central hill region. The *thekthiti* system was used in the Baisi region, as well as in such peripheral districts as Rolpa, Pyuthan and Salyan and in Pallokirat, in the eastern hill region for the collection of taxes on both rice-lands and homesteads.[3] In the Tarai region, a variation of the *thekbandi* system, known as *Panchasala-thek*, had been introduced during the 1820s. In the central hill region, the *amanat*, *ijara*, and *thekbandi* systems were experimented repeatedly for the collection of homestead taxes in the village during the nineteenth century. Apparently, the objective of the government was not to apply a uniform system on a regional basis in the name of reform, but to introduce appropriate arrangements according to the exigencies of each specific situation at any given time. The sequence of transition from any one of these systems to the other, was, therefore, synchronous, rather than consecutive. That is to say, all these systems existed at the same time in different parts of these regions, and at times, even in different *jagirs* in the same district. Nor was the transition invariably one way. There are numerous instances of revenue-collection under the *ijara* system having been replaced by *amanat*, but the number of cases involving a converse process, that is, of transition from *amanat* to *ijara*, is also quite large.

(1) *The Amanat System*[3]

Under the *amanat* system, the government, or the *jagirdar*, appointed a functionary to collect homestead taxes and administer justice in each village. This functionary was known as a *dware*.[4] The *dware* transmitted the actual amount collected to the government or the *jagirdar*, and submitted accounts of collections at the end of each year. The amount of revenue actually collected fluctuated from year to year, hence *amanat* collection did not guarantee a stable income to the *jagirdar*. For the government, the *amanat* system was unsatisfactory because it could not capitalize on judicial fines and extra-legal perquisites to raise the official assignment value of lands and villages.

(2) *The Thekbandi System*

Because nineteenth century official documents usually use the terms *thekbandi* and *thekthiti* as if these were interchangeable, it appears necessary to give them specific definitions for the purposes of the present study. *Thekbandi* may be defined as a settlement with *mukhiyas* in their individual capacity for the collection of revenue for a specific period. When the settlement was made on a long-term basis with the villaje community as a whole represented by the *mukhiya*, the system was known as *thekthiti*.

Under the *thekbandi* system, the amount of taxes and levies due on homesteads actually existing at the time of the settlement was first ascertained. To that figure was added the estimated income from *asmani* payments, usually at a specified percentage of the homestead-tax assessment. The total amount thus assessed was payable on a contractual basis, with no remissions for depopulation, and no increment for new homesteads. However, the government usually reserved the right to appropriate additional income from special levies, expiation fees, and fines collected from persons guilty of serious crimes. *Thekbandi* settlements were usually made with local *mukhiyas*. For an example of *thekbandi*, we may refer to the settlement made in Upallo-Ghachok village in Kaski district during 1837. The village consisted of 90 homesteads. Payments

assessed on these homesteads totalled Rs 104/5,[5] inclusive of Rs
66/4 from the *serma* tax and other sources, half of that amount, or
Rs 33/2, from *asmani*, and Rs 4/15 from the *saunefagu* tax. In
consideration of his contractual liability for the payment of this
amount, the *mukhiya* was allowed to collect and appropriate the
proceeds of taxes from all sources other than those reserved by the
government, as well as *asmani* income.[6]

The *thekbandi* system was introduced in the Tarai districts also
during the late 1820s for the collection of land and other revenues
in a modified form known as *panchasala-thek*. Under that system,
each local *chaudhari* stipulated the amount of revenue for the
parganna under his jurisdiction. The amount was payable on a
contractual basis, so that remissions were allowed only in exceptional circumstances, such as widespread drought. Settlements
under the system were made for five years at a time.[7]

(3) *The Thekthiti System*

Under the *thekthiti* system, the village community as a whole,
represented by the *mukhiya*, and not the *mukhiya* in his individual
capacity, was held liable for the full payment of the revenue. In
matters relating to the assessment and collection of taxes in the
Baisi region, the government dealt not with individual peasants,
but with the community as a whole. The entire village was treated
as one unit for purposes of taxation, leaving it to the headman to
apportion individual shares of the total revenue assessment.[8]
Repeated efforts had been made during the early years of the
nineteenth century to abolish this system, and to assess taxes on
individual holdings as in the central hill region. The government
eventually realized that "systems applicable to other parts of the
country cannot be applied to this region," and, therefore, restored
"the traditional systems." At the same time, it tried to ensure that
the volume of revenue assessment did not decline.[9]

The *thekthiti* system was thus a *modus vivendi* between the traditional system of treating the entire village as a single unit of taxation and the government's desire to establish a correlation between
the amount of revenue assessed and the actual number of homesteads. Under that system, each village was allowed to retain the
proceeds of taxes on both rice-lands and homesteads, as well as
income from duties levied on horses, falcons, wax, printing of
cloth, ferry services, judicial fines and escheats, and one-fifth of

the income from musk exports. Taxes on newly-reclaimed lands were collected in addition to the revenue stipulated from each village. If the amount that was actually collected fell short of the figure assessed, each local household made equal contributions to meet the loss. On the other hand, if actual collections were in excess of the stipulated amount, the surplus was apportioned equally among each local household.[10]

Both the *thekbandi* and the *thekthiti* systems contained a built-in mechanism to protect the government against loss of revenue. Liability to meet shortfalls, if any, was borne either by the entire local community represented by the *mukhiya* under *thekthiti*, or by the *mukhiya* himself under the *thekbandi* system. Settlement orders under the *thekbandi* system declared, as in the Upallo-Ghachok village of Kaski district in 1837:[11]

> *Ryots* shall not make any payment in excess of this amount. (The *jagirdar*) shall not make any additional collection. If he does so, report the matter to us. Promote land reclamation and settlement, and make the village populous. (The *jagirdar*) shall not demand anything for newly-created households, nor, shall you be entitled to remissions for depopulated households.

The *thekthiti* system was particularly appropriate for outlying and inaccessible areas in the hill regions where frequent revenue settlements were not feasible. A decline in the local population caused no loss of revenue to the government, inasmuch as the loss was automatically shared by the remaining households.[12] The converse was also true, however. The amount of revenue under the *thekthiti* system remained unchanged during the interval between two revenue settlements. Consequently, revenue collections remained static for long periods of time. During the half-century between 1837 and 1888, for instance, revenue collection in Rolpa district increased by only ten per cent.[13]

Similarly, under the *panchasala-thek* system followed in the eastern Tarai region, the government could claim only the amount stipulated by them during a five-year period. Extension of the cultivated area and increase in population during that period, therefore, benefitted *chaudharis*, not the government. A system that provided no scope for increasing revenue receipts in the context of the progressive extension of the cultivated area in the Tarai region during the nineteenth century was naturally bound to be un-satisfactory for the government.

(4) *The Ijara System*

The foregoing sections showed how homestead taxes in the hill regions were collected either through *dwares* under the *amanat* system, or through *mukhiyas* under the *thekbandi* and *thekthiti* systems. Under both these systems, the tax-collecting functionary exercised authority to dispense justice, hence he was also able to exact unauthorized payments and labour services from the villagers. The benefits of these extortions and perquisites were substantial enough to attract offers of higher payments from speculators or *ijaradars*. The government could not withstand the temptation of maximizing its revenue by accepting such offers. It, therefore, occasionally bypassed the *mukhiya* and entrusted the responsibilities of homestead tax collection to any person who stipulated a higher amount of revenue. The *mukhiya* was then faced with two alternatives: he could either match the offer, or quit.[14] *Ijara*, therefore, denoted a system under which authority to collect specific taxes or levies, or revenues from all prescribed sources in a specified area, was granted to an individual, or *ijaradar*, who undertook all risks of fluctuations in receipts and stipulated the payment of a fixed sum of money to the government on a periodic basis. The *ijaradar* was allowed no remission if he was unable to make collections for any reason, even because of circumstances beyond his control. At the same time, he was allowed to appropriate any amount that he could collect in excess of the payment he had stipulated.[15]

Ijara, no doubt, augmented the revenue, but subjected the peasantry to harassment and extortion. Indeed, the chief objective of the *ijara* system was to institutionalize and capitalize on these powers of extortion and harassment that tax-collecting functionaries were able to exercise with impunity. Thanks to this system, the government was able to maximize its revenue from homestead taxation to a level which bore little relationship with the official rates of taxes or the number of homesteads on which these were assessed.

(5) *Mukhiyabhar and Lokabhar Systems*

The land-tax-collection systems followed in the hill regions dur-

ing the early years of the nineteenth century more or less effectively
fulfilled the objective of maximizing revenue. At the same time,
these systems gave due consideration to the exigencies of local historical and geographical considerations. But though these systems
served the needs of the government and the landowning elite, they
took little account of the difficulties and hardships of the people.
This was true particularly of the *ijara* system. Rana policy in the
field of tax-collection, consequently, aimed at the amelioration of
the difficulties and hardships that the *ijara* system caused to peasants in the hill region. To fulfil these objectives, the Rana government introduced the *mukhiyabhar* and *lokabhar* systems in that
region.

Mukhiyabhar meant a system under which a *mukhiya* was allowed to match an *ijaradar*'s offer in his personal capacity. As noted
previously, whenever any other person offered a higher amount of
homestead-tax revenue for any village than what the local *mukhiya*
was paying at the time under the *thekbandi* system, the latter was
required either to match the offer or quit. If an increase in revenue
could not be avoided because of the higher amount of revenue offered by a prospective *ijaradar*, the *mukhiya* could at least ensure
that no *ijaradar* was appointed to collect revenue and exercise
judicial authority in the village. This he was allowed to do by
matching the *ijaradar*'s offer under the *mukhiyabhar* system.[16] The
lokabhar system resembled the *mukhiyabhar* system in many respects, except that initiative for removing the *ijaradar* came not
from the *mukhiya* but from the local peasant community. The
1854 legal code contained provisions under which the local inhabitants could stipulate payment of the amount offered by a prospective *ijaradar* through a representative designated by them for that
purpose. The main condition on which *lokabhar* arrangements
were sanctioned was that liability for any shortfall in collections,
or for defalcations by the representative, should be borne by the
entire community.[17] The *mukhiyabhar* and *lokabhar* systems thus
provided the government with the financial benefits of the *ijara*
system, while freeing the local people from the oppressive burden
of an *ijaradar*.[18]

(6) *The Jimidari System*

The introduction of the *jimidari* system in the Tarai region during the early years of Rana rule was a measure of far-reaching im-

portance in the context of tax collection and agricultural development during the nineteenth century. The reform left unchanged the
existing revenue administrative set-up at the *parganna* level, but
made the *mouja* the primary unit of revenue administration.[19] Functionaries, known as *jimidars*,[20] were appointed in each *mouja* to
collect taxes and promote land reclamation and settlement. The
essence of the new system was the personal liability of the *jimidar*
for the full collection of land and other taxes in the *mouja* under
his jurisdiction, even if lands remained uncultivated for any reason.[21]

What were the circumstances that necessitated the appointment
of *jimidars* to discharge functions pertaining to land and revenue
along with the *parganna*-level *chaudharis* in the eastern Tarai region? There is evidence to suggest that the *panchashala-thek* system,
under which *chaudharis* had been employed to collect taxes in that
region, suffered from a number of defects. The *parganna*, which
comprised a number of villages, proved unwieldy as a unit of revenue administration, with the result that *chaudharis* were unable to
discharge their tax-collection functions satisfactorily.[22] The situation was, in fact, so serious that during the 1850s the government
was able to collect less than two-thirds of land and other tax-assess-
ments in the eastern Tarai region.[23] Nor was this all. The development of the Tarai region was one of the main objectives of Rana
economic policy. Such development necessitated institutional
arrangements to undertake and finance colonization schemes. The
Rana government obviously found it necessary to create a new
class of landholding interests at the local level which would have
sufficient incentive to provide such entrepreneurial talent and investment capital.

The role of the *jimidar* as an agricultural entrepreneur will be
discussed in Chapter 9. In the present context, it appears sufficient
to stress the contribution that the *jimidari* system made toward
the emergence of a new elite group in the Tarai region. As mentioned previously, the area that was placed under the control of a
jimidar was called the *mouja*. A *mouja* comprised lands of two
categories: the peasants' allotments and the *jimidar*'s demesne, called
jirayat.[24] These lands were cultivated through the unpaid labour
and implements of peasants who held allotments in the *mouja*.[25]
Jimidars were, consequently, able to combine the administrative
function of tax-collection with a number of economic privileges

vis-a-vis the peasantry.

Because the agrarian policies of the Ranas had different objectives in the hill region and the Tarai, the *jimidari* system comprised several features that were lacking in the tax-collection systems followed in the former. Firstly, the *jimidar* had no authority to dispense justice, unlike *mukhiyas* and *ijaradars* in the hill regions. Secondly, *jimidari* rights could be sold,[26] but there is no evidence that the position of a *mukhiya* in the hill region was salable during the nineteenth century. Finally, the position of a *jimidar* could not be outbidden by a prospective *ijaradar*. The reasons for these differences can be easily explained. In the hill region, the objective of the Ranas was to garner the maximum possible financial resources from an agrarian economy which possibly had reached the optimum limit of growth and even become virtually stagnant. On the other hand, their objective in the Tarai was to create a new group of agrarian interests which would bear the responsibility of tax collection and simultaneously provide the institutional framework for the mobilization of entrepreneurial ability in order to exploit the region's agricultural development potential. That objective was fulfilled to a considerable extent through the *jimidari* system.

The Status of the Village Elite

The foregoing sections contained a general survey of the different systems followed in different parts of the Kingdom for the collection of land taxes during the nineteenth century: the *amanat*, *ijara*, and *thekbandi* systems in the central hill region, the *thekthiti* system in the Baisi region and Pallokirat in the eastern hill region, and the *dhokre* system for the collection of rice-land taxes. These sections also described innovations made in this field by the Rana government after 1846 through the introduction of the *mukhiyabhar* and *lokabhar* systems in the hill regions, and the *jimidari* system in the Tarai. We shall now make an attempt to analyze the general characteristics of the village elite groups and their impact on the economic life of the peasant.

The most striking characteristic is the existence of separate functionaries for the collection of taxes of separate categories. We have seen how in the central hill region taxes on homestead lands were collected by *dwares*, *mukhiyas*, or *ijaradars*, and on rice lands by *dhokres* and *jimmawals*. There also existed a large number of functionaries at subordinate levels to help *mukhiyas* and *jimidars* discharge their functions. It is important to note that the burden of maintaining this multi-layered hierarchy of functionaries was ultimately borne by the peasant.[27]

Rajas, *birta*-owners, and *jagirdars*, no doubt, occupied the topmost position in the hierarchy of landed interests, but it was *mukhiyas*, *ijaradars*, and *jimidars* who functioned as the immediate overlords of the peasant. For the peasantry, *mukhiyas, ijaradars*, and *jimidars* were representatives not only of their ascriptive landowners, but also of the central government. *Mukhiyas* in the hill regions, in particular, combined police, judicial and administrative powers with their normal function of revenue collection. The extent of the *mukhiya*'s powers was naturally greater under the *thekthiti* system than under the *thekbandi* system, inasmuch as he not only collected land and other taxes but also recruited the labour required to repair and maintain fords, ferries, irrigation channels, forts, and tracks, procured supplies for the army, arrested people who came without valid passports, and closed prohibited tracks.[28] Under both these systems, the *mukhiya* constituted the pivot of the local administrative set-up. His traditional authority over the village community was buttressed by the new status he now enjoyed under the royal seal.

The primary objective of all these systems was to maximize the collection of revenue. The contractual element was common to all the systems that were most widely practised : *ijara, thekbandi, thekthiti* and *jimidari*. The *ijaradar, mukhiya* or *jimidar* paid to the government the stipulated amount; anything that he could exact in addition he could keep for himself. Because of the broad authority vested in these elements, there was seldom any check on what they actually exacted. The only problem for these functionaries was not to squeeze the peasant so hard that he might leave the land uncultivated, or risk the anger of his immediate overlord to approach Kathmandu directly with his complaint.

If these systems were clearly exploitative, there is evidence that the government actually intended them to be so. This is indicated by the absence of any provision to reimburse the village elites adequately for their fiscal and administrative services. *Mukhiyas* in most parts of the country were entitled to no formal emoluments;[29] only in some parts of the Baisi region where the *thekthiti* system was prevalent were *mukhiyas* usually allowed a commission amounting to $2\frac{1}{2}$ per cent of the amount actually collected by them.[30] However, because of the small amounts they collected, these commissions seldom exceeded five rupees each.[31] In these circumstances, it is not surprising that Kathmandu was flooded with petitions from villages in almost all parts of the country complaining against the collection of unauthorized fines and payments by *mukhiyas* and other tax-collection functionaries.

The exaction of unauthorized payments from the peasantry in this manner was, however, only one of several ways in which tax collection functionaries provided insurance for the effort and risk they undertook. *Dhokres* who collected rents on rice lands on behalf of *jagirdars*, for instance, had to undergo considerable effort and risk while purchasing *tirjas* from *jagirdars* and collecting rents from the peasants.[32] The *dhokre* had to make a substantial profit in order to compensate for this effort and risk. A part of that profit came from the *jagirdar*'s pocket, when *tirjas* sold at a discount because of difficulties of collection and other factors. At the same time, the *dhokre* used the authority conferred on him by the *tirja* to demand higher and additional payments from the tenants, and harass him in other ways.[33] The *dhokre* thus appropriated profits at the expense of both the *jagirdar* and the cultivator. For the *jagirdar*, of course, the *dhokre* was a necessary evil. For the cultivator, on the other hand, the *dhokre* was an unmitigated evil, inasmuch as he meant one more layer of authority that had to be kept satisfied.

Defects of the Ijara and Amanat Systems

There is enough evidence to show that the economic burden of unauthorized payments and services was highest when taxes were collected by non-resident persons under the *ijara* system. The appointment of an *ijaradar* for the collection of homestead taxes created several difficulties for the local people. So long as the village *mukhiya* discharged that function, greed for revenues was tempered by deference to local custom and tradition. On the other hand, an *ijaradar*, who seldom had such inhibitions, tried his best to maximize his profits within the short period of time available to him. The *ijaradar* not only collected the amount stipulated by him, but also made a profit for himself. The local people had thus to pay much more than what actually reached the government, the excess constituting the profits of the *ijaradar*.

The disruptive impact that the *ijara* system had on village life and customs when local *mukhiyas* were set aside in favour of *ijaradars* is well illustrated by the experience which the people of the Kagbeni-Barhagaun region in Mustang underwent during the 1850s. Before 1857, taxes were collected in that region by local *mukhiyas*. In that year, however, a non-resident person made a higher offer, and was appointed as an *ijaradar*. The local *mukhiyas*, consequently, lost their customary judicial and revenue functions. The *ijaradar* even replaced them by his own men, exacted unpaid labour, fines, and unauthorized payments for the villagers, and harassed them in several other ways. The villagers approached Kathmandu for the redress of their grievances. The government, however, only reconfirmed the authority of the *ijaradar* subject to the traditional customs and usages of the local inhabitants.[34] It was careful not to disturb the contractual arrangements that it had made with the *ijaradar*, inasmuch as this could reduce the amount of revenue that the latter had stipulated.

In an attempt to resolve the conflict between village *mukhiyas* and non-resident *ijaradars*, the government decreed that *ijaradars* should collect revenues only through *mukhiyas*,[35] not establish direct contacts with the local inhabitants,[36] and not interfere in the judicial functions of the *mukhiya*. The *ijaradar* was empowered, however, to receive complaints against the *mukhiya*,[37] an authority which further disrupted whatsoever cohesiveness the village community originally possessed.

This attempt to achieve a compromise between the traditional authority of the village *mukhiyas* and the rights of the *ijaradar* did not reduce the tax burden on the people. The *ijaradar* was, no doubt, obliged to deal with the villagers only through the *mukhiya*, but he still possessed the authority to collect a sum which was much larger than what the village had been paying customarily. People thus suffered from two disabilities: they not only paid higher taxes, but also lived under the overlordship of a non-resident person who was not bound in any way by local customs and had little interest in their prosperity.

Nor was the condition of the peasant any better under the *amanat* system. The appointment of a *dware* as both bailiff and magistrate created new problems for the peasant. The *dware* was entitled to certain perquisites, but the village *mukhiya*, although he had no place in the new set-up, still wielded enough authority to be able to demand, and receive, his own traditional fees and servitudes. Abuse of his authority by the *dware* added to the difficulties faced by the peasant in thus living under two masters. There is evidence that *dwares* often harassed the villagers in different ways, ignored local customs and usages, and wrongfully evicted people from their lands and homesteads.[38] In 1816, for instance, the *amanat* system was abolished in Sarangkot division of Kaski district when the government received reports that "the *dware* has encroached upon forests and pastures, collected unauthorized payments, exacted unpaid labour, imposed fines for offences which had not been actually committed, and thus oppressed the villagers."[39] *Jagirdars* were, therefore, often ordered not to appoint *dwares*, or dismiss village *mukhiyas*.[40]

Impact on the Agrarian Structure

Systems of land-tax collection in which tax-collection authority was vested on a permanent basis in a local functionary, such as the *mukhiya* and the *jimidar* had also a profound impact on the agrarian structure in the village. There exists a large body of evidence to show that these functionaries often abused their powers of land allotment to grab rice lands and homesteads for themselves. In its desire to prevent revenue from declining, the government often followed policies which encouraged such concentration of land-ownership in the hands of the village elite. For instance, *mukhiyas* were under obligation to allot vacant holdings to landless peasants in the first place. If such peasants were not immediately available, they were permitted to appropriate the holdings for their own use on a provisional basis.[41] However, *mukhiyas* seldom lost such easy opportunities to augment their own holdings and often retained possession of vacant holdings on a permanent basis.[42] Prospective allotees were seldom able to have their legal claims to the vacant holdings upheld against the village elite.

These systems also often unduly depressed the status of the mass of the peasantry. Under the *thekthiti* system, for instance, the peasant was tied down to the fiscal obligations that the possession of a homestead entailed. He was even denied the right to break free from these obligations. Peasants who shifted to other areas were brought back to their homesteads, often through the use of force.[43] At times, they were permitted to go elsewhere only after they stipulated that they would continue paying the taxes assessed on their holdings.[44]

The introduction of the *jimidari* system, in particular, led to an increased degree of polarization in the agrarian community. All classes of peasants were then placed under the jurisdiction of *jimidars*, thereby depressing the status of those independent farmers who had received land allotments directly from the local administration. The tenurial rights of these farmers remained more or less unaffected, but they were placed under the fiscal and administrative authority of the *jimidar*, and obliged to provide labour for the cultivation of his demesne. The *jimidar* had power to evict them if they defaulted in the payment of taxes, and to reallot their holdings to other persons.

To conclude: there seems little doubt that the village elite fulfilled an essential intermediary role in the coalition of interests between the aristocratic and bureaucratic groups of *rajas, birta*-owners and *jagirdars*, or the central government, to extract agricultural surpluses from the peasantry. Each side needed the other. The landowning elite needed the village elite to collect rents and taxes and control the peasantry, and the village elite needed the political backing provided by the landowning elite. The village elite, even though some of them might have had an ancient origin as the focal point of local leadership under traditional systems of communal autonomy, thus gradually assumed the role of representatives of the central government through which the landowning elite tightened their economic stranglehold on the peasantry. It is doubtful if the peasantry derived any benefit from these village elites in their day-to-day problems of eking out a subsistence from their land. The imposition of an elite group on the local agrarian community for the benefit of *rajas, birta*-owners, and *jagirdars* compelled the peasant to bear the costs of his own political and economic domination.

NOTES

[1]This section is based on Mahesh C. Regmi, *Land Tenure and Taxation in Nepal*, Berkeley, University of California Press, 1963-68, Vol. 1, pp. 126-36, and Vol. 3, pp. 13 and 36-38.

[2]"Revenue Functionaries in the Eastern Tarai Districts," *Regmi Research Series*, Year 2, No. 5, May 1, 1970, pp. 107-9.

[3]Scholars who have read a description of the *amanat* and *ijara* systems as followed in the eastern Tarai region during the latter part of the eighteenth century and the early nineteenth century in the author's *A Study in Nepali Economic History, 1768-1846* (New Delhi, Manjusri Publishing House, 1971) will note the difference in the explanation of the same terms in the present study. The difference is due to the fact that the *ijara* and *amanat* systems were introduced at the district level in the Tarai region, whereas the present study is concerned with the use of these systems in the villages of the hill region.

[4]Cf. "Appointment of Shiva as Dware in Tokha," Poush Sudi 13, 1894 (January 1838), *RRC*, Vol. 35, p, 501; Brian H. Hodgson, "Some Account of the Systems of Law and Policy as recognized in the State of Nepal," *Journal of the Royal Asiatic Society of Great Britain and Ireland*, Vol. 1, 1834, pp. 274-75.

[5]The second figure refers to *annas*. The standard Nepali rupee then comprised 16 *annas*. There were also local units of 13, 20, or 24 *anna* rupees.

[6]"Thekbandi Grant for Revenue Collection in Upallo-Ghachok Village, Kaski District," Baisakh Sudi 14, 1894 (May 1837), *RRC*, Vol. 34, pp. 295-96.

[7]Regmi, *Nepali Economic History, op. cit.*, p. 175.

[8]"Jumla Administrative Regulations, 1796," sec. 11, *ibid.*, p. 216; "Order to Chautariya Pushkar Shah regarding Land-Tax Assessments in Pari-Pilkot Village, Doti District," Jestha Badi 11, 1889 (May 1832), *RRC*, Vol. 27, p. 135.

[9]"Royal Order to the Subba of Jumla," (Baisakh Sudi 12, 1864 May 1807), in Yogi Naraharinath (ed.), *Itihas Prakash* (Light on History), Kathmandu, Itihas Prakash Sangh, 2013 (1956), Vol. 2, pt. 2, p. 16; "Royal Order to the Thanis, Tharis, and Ryots of Jumla," Baisakh Sudi 12, 1864 (May 1807), *RRC*, Vol. 20, p. 291.

[10]"Regulations regarding Thek-Thiti in Jumla," Chaitra Sudi 7, 1900 (March 1844) *ibid.* Vol. 34, pp. 618-24; "Thek-Thiti Arrangements in Jumla," Falgun Sudi 3, 1894 (February 1838), *ibid.* Vol. 35, pp. 566-601; "Thek-Thiti Arrangements in Chainpur," Aswin Sudi 2, 1904 (September 1847), *ibid.*, pp. 389-93.

[11]"Thek-Thiti Grant for Revenue Collection in Upallo-Ghachok Village, Kaski District," 1837. (See n. 22 above).

[12]"Order regarding Thek-Thiti Revenue Collection in Rukum"; Kartik Badi 5, 1943 (October 1886), *RRC*, Vol. 51, pp. 871-77.

[13]"Reconfirmation of Thek-Thiti System in Rolpa," Kartik Sudi 15, 1945 (November 1888), *RRC*, Vol. 53, pp. 732-49.

[14]"Reinstatement of Motiram Thapa Chhetri as Mukhiya in Gotibad Village, Pyuthan District," Baisakh Sudi 1, 1919 (April 1862), *RRC*, Vol. 29, pp. 471-72.

[15]Cf., "Ijara Grant to Simhabir Thapa for Revenue Collection in Tinpatan," Shrawan Sudi 4, 1895 (July 1838), *ibid.* Vol. 35, pp. 553-55.

[16]For instance, in four villages of the Chharka-Bhot area in modern Dolpa district, which had been assigned as *Jagir*, homestead taxes had been assessed under the *Thek-Thiti* system in 1836. These taxes were collected from 1860 to 1863 by *ijaradars* on payment of Rs 9,648 yearly. In 1864, the local *mukhiyas* stipulated payment of this amount themselves. The government accepted their request, but on the condition that the *mukhiyas* should match the higher amount of revenue if stipulated by any person at any time in the future. "Mukhiyabhar-Thek Grant in Debung and Other Villages in Chharka-Bhot," Magh Sudi 15,

1921 (February 1865), *ibid.*, Vol. 49, pp. 450-61.

[17]Government of Nepal, "Rairakamko" (On Revenue Matters) sec. 3, in Ministry of Law and Justice, *Shri 5 Surendra Bikram Shahdevaka Shasanakalma Baneko Muluki Ain* (Legal code enacted during the reign of King Surendra Bikram Shah Dev), Kathmandu, the Ministry, 2022 (1965), pp. 52-53; "Lokabhar Grant in Villages of Tilpung," Shrawan Sudi 1, 1914 (July 1867), *RRC*, Vol. 49, pp. 323-27.

[18]In Khinchet and some other villages of Nuwakot district, for instance, *mukhiyas* were paying Rs 1,761 on *thekbandi* basis every year during the early 1860s. A prospective *ijaradar* offered Rs 2,461, but his offer was rejected when the *mukhiyas* stipulated Rs 2,161 under the *mukhiyabhar* system. The *mukhiyas* were thus burdened with an additional payment of Rs 400 because the government had received a higher offer than what they had been paying previously. The government, on its part, was able to make a show of generosity by accepting Rs 300 less than what it had been offered by a third party. "Order regarding Mukhiyabhar Arrangements in Khinchet," Jestha Sudi 4, 1923 (May 1867), *ibid.*, Vol. 57, pp. 464-70.

[19]The term *mouja* is of Arabic origin, which Nepal has obviously borrowed through India. The term, as used in India, denoted "a considerable area of cultivated and waste land (which may contain several villages) aggregated for the purpose of record and revenue collection." B.H. Baden-Powell, *The Land Systems of British India*, (reprint of 1892 ed.), Delhi, Oriental Publishers, n.d., Vol. 3, p. 420.

[20]The term *Jimidar* should not be confused with *Zamindar*, or village landlord. Cf. Mahesh C. Regmi, *Landownership in Nepal*, Berkeley, Los Angeles and London, University of California Press, 1976, pp. 106-7. See also Chapter 7.

[21]"Revenue Regulations for the Eastern Tarai," Marga Badi 6, 1918 (November 1861), *RRC*, secs. 23 and 42, Vol. 10, pp. 18-19 and 30. These regulations provide no role to the *Chaudhari* in the collection of revenue. In the far-western Tarai, on the other hand, both the *Chaudhari* and the *Jimidar* were jointly responsible for the discharge of that function. According to revenue regulations promulgated for that region in 1860: "In case the *Jimidar* dies or escapes to India, or is unable to complete payments within a period of one month after the expiry of the year, the *Chaudhari* may auction his property to recover the arrears. Arrears for the entire *parganna* shall be realized from the *Chaudhari* .... In case arrears cannot be recovered in this manner, or in case the *Chaudhari* dies or escapes to India, his property shall be auctioned, and the arrears shall be recovered from the proceeds of the auction." "Revenue Regulations for the Naya Muluk Region," Kartik Sudi 15, 1917 (November 1860), *ibid.*, Sec. 33, Vol. 47, p. 433.

[22]In 1848 Kathmandu received reports that in Saptari district "taxes are not being collected in a satisfactory manner, nor are accounts being cleared in time." "Royal Order to the Chaudharis of Saptari," Jestha Sudi 2, 1905 (May 1848), *ibid.*, Vol. 33, p. 70.

[23]In 1852, actual collection of land taxes, market duties, export duties on timber, and pasturage taxes in the eastern Tarai districts totalled Rs 651,563

(68.0 per cent), against an assessment of Rs 1,044,522. The corresponding figures for 1858 were Rs 567,723 and Rs 1,390, 820 (40.8 per cent). These statistics have been compiled from revenue records of the government of Nepal for the eastern Tarai districts for the relevant years.

²⁴In the eastern Tarai region, the *Jimidar* was granted on a taxable basis cultivated lands yielding an income amounting to 5 per cent of the total amount of revenue assessed in the *mouja* under his jurisdiction. "Revenue Regulations for the Eastern Tarai," Marga Badi 6, 1918 (November 1861), *RRC*, Sec. 58, Vol. 10, pp. 44-45. The percentage was 10 in the far-western Tarai region. "Revenue Regulations for the Naya Muluk Region, "Marga Badi 6, 1918 (November 1861), *ibid.*, Sec. 62, Vol. 47, p. 467. A system of land allotments to individual peasants and the peasants' obligation to work on the demesne farms of the lord were thus essential characteristics of both the *jimidari* system and the feudal system of medieval Europe. Nevertheless, it will be erroneous to deduce from these similarities that the *jimidari* system was essentially feudal in character. Feudalism, in the classical European sense, implies serfdom, a condition in which the peasant does not own the land he cultivates, but is forced to cultivate it in order to fulfil the economic demands of the lord. In contrast, the allotment of a peasant in the Tarai region during the nineteenth century was legally recognized as his personal property which he was free to sell, bequeath, or relinquish.

²⁵"Revenue Regulations for the Eastern Tarai," Marga Badi 6, 1918 (November 1861), *RRC*, Sec. 28, Vol. 10, p. 21.

²⁶"Order regarding Transfer of Jimidari Holdings in Bardiya," Ashadh Sudi 4, 1854 (June 1897), *ibid.*, Vol. 61, pp. 93-106.

²⁷In Morang district, for example: "Each village (Ganj) is under a Mokuddum, who has five *per cent* of the land free of rent; a *patwari* or clerk, who has one-half anna on the rupee of rent, and two annas a year on each house, both from the tenants, and gives one- half of this to his superiors. . . . The messengers (Gorayits) from every house get about two loads of the ears of rice, which give about one *man* (82 lbs. avoirdupois) or grain." Francis Hamilton, *An Account of the Kingdom of Nepal* (reprint of 1819 ed.)New Delhi, Manjusri Publishing House, 1971, p. 155. The post of *Mokuddum* appears to have been abolished after the introduction of the *Jimidari* system, but the other functionaries were retained. Cf. Regmi, *Land Tenure and Taxation, op. cit.*, Vol. 1, pp. 134-35.

²⁸Cf. "Thek-Thiti Arrangements with Subba Sridal in Chainpur," Baisakh Badi 30, 1901 (May 1847), *RRC*, Vol. 35, pp. 274-76; "Thek-Thiti Arrangements with Mukhiyas in Chingadh, Dailekh District," Marga Badi 5, 1938 (November 1881), *ibid.*, Vol. 58, pp. 129-40.

²⁹Cf., "Order regarding Thekbandi Arrangements in Salyan," Jestha Sudi 14, 1935 (June 1878), *ibid.*, Vol. 50, pp. 61-66.

³⁰"Order regarding Thek-Thiti Arrangements in Mugu and Elsewhere in Jumla District," Marga Badi 8, 1937 (November 1880), *ibid.*, Vol. 58, pp. 61-74.

³¹Cf., "Order regarding Remuneration of Mukhiyas in Doti," Falgun Sudi 9, 1922 (March 1866), *ibid.*, Vol. 49, pp. 592-98.

³²For instance, if crops were damaged by natural calamities after *Dhokres* purchased these certificates, they were required by law to grant remissions to

the cultivator. However, they were not permitted to claim any refundment from the *Jagirdar* to compensate the loss. Often *Jagir* lands were located at distant places and this added to the difficulties of collection. *Dhokres* were prohibited from making collections in kind if the cultivator preferred to commute his in-kind tax-liability at current market prices. Government of Nepal,"Bali Bikriko" (On the sale of rents), secs. 3-5, in *Shri 5 Surendra . . . Muluki Ain*, pp. 50-51; "Regulations regarding Collection of Rents on Jagir Lands in Salyan and Pyuthan," Baisakh Badi 8, 1954 (April 1897) *RRC*, Vol. 61, pp. 594-603.

³³Regmi, *Land Tenure and Taxation, op. cit.*, Vol. 3, pp. 39-40.

³⁴This accout of tax-collection problems in the Kagbeni-Barhagaun region is based on the following documents: "Royal Order regarding Tax-Collection in Barhagaun," Falgun Badi 6, 1881 (February 1825) *RRC*, Vol. 40, pp. 558-59; "Royal Order regarding Dipute between Bisht, Budhas and Other Inhabitants of Barhagaun," Shrawan Sudi 11, 1884 (July 1827), *ibid.*, Vol. 43, pp. 235-39; "Royal Order to the Amali, Budhas, and other Inhabitants of Kagbeni-Barhagaun," Chaitra Sudi 14, 1950 (April 1864), *ibid.*, Vol. 49, pp. 194-205.

³⁵"Royal Order to Tula Shahi and Shaktimalla Shahi," Ashadh Sudi 13, 1895 (July 1838*)*, *ibid.*, Vol. 35, pp. 486-87.

³⁶"Orders to the West No. 2 Bakyauta." Ashadh Sudi 12, 1948 (July 1891), *ibid.*, Vol. 50, pp. 242-50.

³⁷"Royal Order to the Amali, Budhas, and other Inhabitants of Kagbeni-Barhagaun," April 1864. (See n. 34 above).

³⁸"Royal Order to Dware Sadhuram of Gaikhur in Gorkha District," Marga Badi 14, 1875 (November 1819), *RRC*, Vol. 38, p. 433; "Royal Order regarding Eviction of Kripa Kandel in Chiti, Lamjung District," Chaitra Badi 11, 1875 (March 1819), *ibid.*, p. 550.

³⁹"Order regarding Thekbandi Arrangements in Sarangkot, Kaski District," Bhadra Badi 5, 1873 (August 1816), *ibid.*, Vol. 28, pp. 144-45.

⁴⁰"Order regarding Collection of Rents on Jagir Lands in Areas West of Bheri River," Chaitra Sudi 4, 1934 (April 1828), *ibid.*, Vol. 67, pp. 262-71.

6

Unpaid Labour

The previous chapters described how surplus agricultural production was collected from the peasant in money or commodities in different parts of the Kingdom. These payments did not represent the total burden on the peasantry, because peasants in nineteenth-century Nepal were also compelled to work without wages for the government, for the landowning elites, and, in fact, for any other individual who was placed over them in a position of authority. This chapter will discuss how these unpaid-labour obligations were given an institutional form in nineteenth-century Nepal, and imposed an additional economic burden on the peasant.

Unpaid Labour and the Landowning Elite

The right to exact unpaid labour on a compulsory basis from all classes of inhabitants traditionally belonged to the government. When the lands tilled by a peasant were granted by the government to any individual under the *rajya*, *birta* and *jagir* systems, the peasant's unpaid labour obligations automatically became due to the appropriate *raja*, *birta*-owner, or *jagirdar*, rather than to the government. Little information is available regarding the nature of the unpaid-labour services that *rajas*, *birta*-owners, and *jagirdars* exacted from their tenants, and the incidence of the burden such services imposed. The relationship between landlord and peasant is essentially a private one; hence descriptions of its actual operation in official documentation are seldom available. However, contemporary accounts indicate that porterage was one of the avenues in which the landowning elites traditionally utilized their right to exact labour from their tenants. As Kirkpatrick recorded in 1793:[1]

> Persons of a certain rank have suitable establishments of Durwars, or hammocks, without, however, regularly maintaining bearers for the carriage of them, it being among the obligations of the tenants of jaghires and other landed estates, to perform this service occasionally for the proprietor.

There is also evidence that the landowning elites utilized this privilege for the supply of firewood and other household necessities, and for the transportation of rents.[2]

Multiple Obligations

Peasants were under obligation to provide unpaid-labour services not only to their *raja*, *birta*-owner, *jagirdar*, or other landlord, but also to the government.[3] A peasant who cultivated lands belonging to a *birta*-owner thus owed unpaid-labour services both to the government and to the *birta*-owner. This made it possible for such peasants to claim that "even though the lands we cultivate have been granted as *birta*, we remain subjects of His Majesty."[4] At the same time, the burden of such dual obligations may have been somewhat lighter on peasants cultivating *birta* lands, compared with *jagir* and other lands. Kirkpatrick's account is illuminating in this respect also:[5]

> The Ryots, or peasantry, are distinguished into Kohrya and Perjah; the former are those settled in Birta proprietory, or other rent-free lands, and are not liable to be called on by government for any services except the repair of roads, and attendance on the army upon particular occasions. Those Perjahs who occupy lands actually belonging to the Prince, though, perhaps, in the immediate possession of jaghiredars, are, on the contrary, obliged to perform various services, both at the call of the jaghiredar, and of the Prince.

Nor was this all. Chapter 5 had discussed how the landowning elites collected their rents from the peasantry through such local functionaries as *mukhiyas, ijaradars,* and *jimidars.* In most cases, therefore, it was these local functionaries, rather than the landowning elites, who actually exercised the right to exact unpaid labour on a compulsory basis from the peasant. In 1882, for instance, the Raja of Malbara, Narendra Singh Raya, appointed a *jimidar* to collect revenue in the village of Dhanbod in Dullu on his behalf. The privileges he granted to the *jimidar* included the right to exact unpaid labour from the inhabitants of that village.[6] It is easy to explain the reason for such delegation of authority. Labour power cannot be accumulated, and the low level of monetization of the economy made the commutation into money of unpaid labour obligations seldom feasible. The landowning elites, therefore, made a virtue of necessity by delegating to their local agents a right that they could not exercise themselves.

When revenue collection arrangements on *jagir* and other categories of *raikar* lands were made by the government, local functionaries were often forbidden to exact unpaid labour for their own use from the inhabitants of the areas placed under their jurisdiction. Such prohibition formed a part of the revenue-collection arrangements made with *mukhiyas* under the *thek-thiti* system in the Baisi region.[7] It is difficult to believe that *mukhiyas* actually complied with such regulations. In any case, they were responsible for recruiting the unpaid labour required for public purposes,[8] hence it does not seem likely that the ordinary peasant was in a position to refuse to work for them in their private capacity. At the same time, some categories of local functionaries enjoyed a statutory right to the free labour of the local peasantry. For instance, the *jimidars* of the Tarai region were entitled to the free use of one ox-team from each local household, or of three labourers, per year.[9]

The Labour-Tax: Jhara and Rakam

When the peasant was required to work for both the government and his landlord, the government's needs naturally received priority. Indeed, the peasant's obligation to work without wages for his landlord was in addition to a labour-tax traditionally due to the government, and consequently, formed but a small part of his total unpaid-labour obligations. The government's demand for labour was high during the early years of the nineteenth century, particularly for the transportation of mail and military supplies, whereas the economy had not been sufficiently monetized to make possible the payment of wages in money. This left no alternative for the government but to impose a labour-tax on peasant households.

The labour tax usually assumed two forms in nineteenth century Nepal: *jhara* and *rakam. Jhara* meant the casual impressment of unpaid labour for meeting occasional needs. *Rakam,* on the other hand, denoted the imposition of a labour tax on the peasantry for the regular discharge of a specific function.[10] The significance of the distinction between these two forms of the labour tax— *jhara* and *rakam*—lies in the fact that the state exercised authority over the peasant not only as a tenant but also as a subject. The peasant was usually under obligation to provide labour services under the *jhara* system whether or not he cultivated an agricultural holding. *Rakam,* on the other hand, was a part of the peasant's obligations as a tenant. *Rakam* services, therefore, automatically came to an end if the peasant relinquished his holding. These two forms of subjection were seldom interlinked. A peasant who was enrolled under the *rakam* system was usually granted exemption from the obligation to provide *jhara* services.

Jhara labour was exacted in almost all parts of the country during the latter part of the eighteenth century.[11] It was subsequently banned in the eastern Tarai region because its indiscriminate use prevented peasants from cultivating their lands properly, but the ban appears to have been largely ineffective.[12] Indeed, local administrators continued to exact unpaid labour without any restriction in the Tarai region throughout the nineteenth century.[13] Even then, wages were paid in money for many labour services in the Tarai region that were exacted without any payment in the hill region, where the *rakam* system remained confined throughout the nineteenth century.[14]

Categories of Unpaid-Labour Services

In order to determine the incidence of labour taxation, it is necessary to describe the different categories of unpaid labour services that were impressed from the peasantry under the *jhara* and *rakam* systems. An exhaustive enumeration of these services would be neither feasible nor meaningful, inasmuch as the government traditionally exercised the right to exact such services for any purpose whatsoever. Broadly speaking, *jhara* labour was utilized chiefly for the occasional construction and repair of roads, bridges,

irrigation channels, and other public utilities. It was also customarily utilized for such miscellaneous purposes as the supply of fod-
der to the royal stables, or the cultivation of lands assigned for the
supply of foodgrains and other provisions for the royal kitchen.
The *jhara* system, moreover, made it possible for the royal palace
to obtain a regular supply of mangoes from the Tarai, and of ice
in the summer months from the hill areas around Kathmandu
Valley.[15]

Inasmuch as *jhara* was a manifestation of the peasant's personal
subjection to the state, there was no limit to the scale on which
unpaid labour services could be exacted under that system. On the
other hand, the government's demand for labour services was limited by its administrative, military, and other needs. Available
supplies of *jhara* labour were, consequently, always in excess of
the actual demand. Moreover, often it was not feasible to impress
jhara labour in outlying areas because of problems connected with
communications and the maintenance of law and order. Mainly
because of these reasons, labour-tax obligations under the *jhara*
system were commuted into a cash levy in many parts of the hill
regions during the early years of the nineteenth century.[16] The
commutation of *jhara* obligations, however, did not liberate the
peasantry from the obligation to provide unpaid labour services
for the essential requirements of the government. Consequently,
commutation often only provided an additional means to increase
the burden of taxation on the peasantry.

Rakam services, on the other hand, were for such regular administrative and defence needs as the transportation of mail and
government supplies, mining and the manufacture of munitions,
and the management of checkposts. These services were utilized
also for the supply of such goods as building timber, firewood and
charcoal, and earthen vessels, and for impressing the services of
masons, carpenters, blacksmiths, stone-cutters, bricklayers, and
gardeners for meeting the personal and household needs of the
royal family. The relative importance of various categories of *rakam*
services differed in Kathmandu Valley and other parts of the hill
region. The central departments of the government, the households
of members of the royal family and, later, the Rana family, military installations, and several munitions factories were located
within Kathmandu Valley. *Rakam* services concerned with the
supply of building timber, firewood, charcoal, etc., and the serv-

ices of artisans and labourers were, therefore, of primary importance in the area. Elsewhere in the hill regions, porterage services
were most extensively used under the *rakam* system, followed by
services in the fields of mining and munitions.

There were two categories of porterage services in the hill regions: *kagate-hulak* for the transportation of official mail, and
thaple-hulak for the transportation of arms and ammunition, gifts
and presents to the royal palace, sick and injured military person-
nel, and "such other goods as may be specified from time to time
under the signature of the Prime Minister."[17] Throughout the
nineteenth century, *hulak* services remained the most important
component of the *rakam* system. These services, in fact, constituted the foundation of the logistics system that enabled the Gorkhali army to fight prolonged campaigns over vast distances.[18] The
importance of these services continued even after the phase of territorial expansion came to an end with the Nepal-British war of
1814-16, because they helped to maintain a line of communication
throughout the length and breadth of the Kingdom.

Rakam services were utilized, as mentioned previously, also for
the production of munitions. Captain Orfeur Cavanagh, who
visited Kathmandu in 1851, mentions a foundry in Kathmandu,
and "a large manufactory of fire-arms" in Pyuthan. He estimated
that in an emergency the government of Nepal could supply "mus-
kets and accoutrements sufficient to equip upwards of 100,000
men."[19] Twenty-six years later, in 1876, another British official,
Sir Richard Temple, observed that "in the Valley near Kathmandu
there are arsenals and magazines, with ordnance, including siege
guns, stores, thousands of stands of arms, small arm ammunition,
and the like." He found it "remarkable" that for all this "they
depend on indigenous manufactures."[20] For unskilled work, as
well as for the supply of materials, these factories depended on
unpaid labour under the *jhara* and *rakam* systems. In Pyuthan district:[21]

People are being employed in different capacities to meet the
requirements of the local munitions factory. In some villages,
people extract iron ore, while others transport the iron to the
factory. Still other people procure and supply timber, charcoal,
hides and skins, saltpeter, sulphur, borax, or salt. People are also
employed to grind gunpowder, or construct factories and other

government buildings, bridges, etc. Other obligations include the supply of stones, flints, sand, wax, baskets, oil, oilcakes, oilseeds, etc. The people of this district have thus to remain in constant attendance at the factory all the twelve months of the year.

The unpaid labour obligations of the inhabitants of villages situated on the Nepal-Tibet borders were equally burdensome, because these involved both porterage and military services. In the Panchsayakhola area of Nuwakot district, for instance, several villages were assigned for the transportation of mail and government supplies, and the rest for work at the local gunpowder factory. In addition, each village was ordered to:[22]

Seize arms and ammunition, saltpeter, sulphur, etc., being smuggled to Tibet, auction them, and transmit the proceeds to the palace. Capture any rebel who may try to escape to Tibet through that area, as well as persons who attempt to create disturbances, and send them to the palace. Provide porterage services for the transportation of government supplies between Nepal and Tibet. Also provide assistance in the collection of customs duties on goods traded between Nepal and Tibet.

Similarly, the checkposts that the government maintained on routes leading toward the south through the Mahabharat mountains were manned by sentries and guards employed under the *rakam* system. Their main functions were to ensure that restricted forests were not cleared, that unauthorized tracks were not used, and that people were not allowed to travel without passports.[23]

The evidence thus seems to be fairly clear that at the middle of the nineteenth century, labour taxation under the *jhara* and *rakam* systems was exacted for a variety of purposes, and on an extensive scale, in the hill regions. Because these services entailed no financial obligation on the government, no attempt appears to have been made at any time to check indiscriminate or wasteful exaction.

Discriminatory Systems

The *rakam* system, had it been carried to its logical conclusion, would have brought the entire peasantry within its scope. However, at no time does the government appear to have found it necessary to mobilize unpaid labour on such a vast scale. Throughout the whole of the nineteenth century, therefore, a large body of peasants in the hill regions remained outside the ambit of the *rakam* system. These peasants were known as *Chuni*. *Chuni* peasants played two essential roles in the labour-tax system of nineteenth-century Nepal. They were the main source of *jhara* labour, because, as mentioned previously, *rakam* workers were usually exempt from *jhara* obligations. Moreover, *Chuni* peasants constituted a reserve labour force for enrolment under the *rakam* system when necessary. The imposition of *rakam* obligations on such a discriminatory basis may have been dictated by the government's actual need for labour services, as well as by the need to maintain a labour force in reserve for *jhara* and future *rakam* impositions. From the viewpoint of the peasantry, however, it made the labour-tax system discriminatory in character.

Discrimination was practised in labour-taxation also on the basis of caste status. During 1813-14, for instance, Brahmans of both Jaisi and Upadhyaya categories were granted exemption from labour-tax obligations under the *jhara* system.[24] The burden of such obligations, consequently, fell more heavily on the non-Brahman sections of the population. Exemption from *jhara*, however, did not mean exemption from *rakam* obligations. Indeed, Brahmans of both Jaisi and Upadhyaya categories seem to have been given preference in enrolment as mail-carriers under the *hulak* system. During the 1850s, for instance, the overwhelming majority of the 4,979 *hulaki* households providing mail transportation services from Kathmandu to Dhankuta in the east and Doti in the west consisted of Jaisi and Upadhyaya Brahmans,[25] whereas the enrolment of members of untouchable communities was usually banned.[26]

Fiscal and Tenurial Facilities

Underlying the distinction between the two categories of labour-tax obligations, *jhara* and *rakam*, was the element of force in *jhara*, and of persuasion in *rakam*. People whose services were impressed for non-recurring functions under the *jhara* system knew that they were fulfilling an obligation traditionally due to the state. They had no choice but to provide these services, because they happened to be present at the appropriate time and place. On the other hand, *rakam* workers performed a recurring function; hence they knew what was expected of them in the future. *Jhara* labour could be rounded up by force, if necessary, but *rakam* obligations needed sustained motivation. The government made an attempt to sustain such motivation through the provision of certain facilities and privileges to *rakam* workers. The benefits that accrued to *rakam* workers from these facilities and privileges partially offset the economic burden of labour-taxation, and hence need to be studied in more detail in the present context.

The main facilities and privileges that most categories of *rakam* workers enjoyed during the early years of the nineteenth century were exemption from the payment of homestead taxes and levies and security of occupancy rights on their rice-lands subject to the payment of rents to their *jagirdar*-landlords.[27] These benefits assume special significance when we note that *jagirdars* at that time had full authority to evict their tenants and relet their lands to any person who offered higher rents.

However, security of occupancy rights was of little significance if the *rakam* worker had no rice lands to cultivate, or if his rice-land holding was too small to provide a subsistence. A policy of providing allotments of rice lands to porters of different categories employed to transport government mail under the *kagate-hulak* system was, therefore, initiated early during the nineteenth century. Arrangements were made in the western hill region to provide each *hulaki* household with a rice-land allotment of between 40 and 100 *muris*, depending on the size of the family.[28] On the eve of the Nepal-British War, these arrangements were extended to *hulaki* porters in the eastern hill region also.[29] During the 1840s, rice-land allotments were similarly provided to several other categories of *rakam* workers in Kathmandu Valley.[30]

These arrangements occasionally led to the transfer of rice-lands to *hulak* porters from other categories of the local peasantry and thereby benefitted one class of the peasantry at the expense of another. To avoid that situation, the government in subsequent years enrolled as *hulak* porters only those peasants who possessed comparatively large rice-land holdings.[31] In other words, peasants were forced to accept labour obligations under the *rakam* system not to gain any additional benefit through rice-land allotments, but only to be allowed to retain the lands that they already possessed.

In any case, rice-land allotments under the *rakam* system were by and large limited to the central hill region; the Baisi region remained outside the ambit of these measures. Differences in the land-tenure systems customarily followed in the Baisi regions and elsewhere in the hills will be discussed in the next chapter; the point that needs to be stressed in the present context is that the state exercised greater control over the allotment of agricultural lands in the central hill region than in the Baisi region. The policy of tying *rakam* obligations with land-holding was, consequently, seldom feasible in the Baisi region. Arrangements were, therefore, made during the early years of the nineteenth century to pay wages in money to *hulak* porters in the Baisi region, but this was merely a temporary expedient during a period of war. The government was not financially capable of bearing the burden of such payments in normal times, hence they were soon discontinued. *Hulak* porters were, therefore, usually compensated through exemption from homestead taxes and security of tenure, without any provision for rice-land allotments.[32]

Abuse of the Labour-Tax System

The foregoing account shows that labour-taxation under the *jhara* and *rakam* systems imposed an additional economic burden on the peasant on a discriminatory basis, although in the case of *rakam* the burden was partially mitigated through fiscal and tenurial concessions. Any attempt to assess that burden would prove to be a mere academic exercise if we ignore the additional burden caused by the abuses that were inherent in the system. It is true that both *jhara* and *rakam* labour was usually utilized for meeting governmental needs. It is also true that *rakam* obligations were imposed only through the orders of the central government in Kathmandu. But it is equally true that the actual exaction of labour services under these systems was the responsibility of village headmen, petty officials, and local administrators. The abuse of *jhara* and *rakam* services for meeting personal needs, in addition to the statutory needs of the government, was thus a built-in defect of the system. To what extent the system was abused cannot, of course, be assessed in quantitative terms, but available evidence suggests that it was grossly and widely abused.[33] The peasant had no protection against such exploitation, and, with stray exceptions, appears to have accepted it docilely. Docility, nevertheless, did not preclude passive resistance, and that was the weapon that the peasants of nineteenth-century Nepal appear to have used on an almost country-wide scale against the abuses of the labour-tax system. Such passive resistance usually assumed the form of voting with their feet. Peasants who found the burden intolerable deserted their homesteads and villages. Orders were promulgated from Kathmandu repeatedly forbidding the exaction of unpaid labour for meeting the personal needs of village headmen and local officials, but these orders were seldom effective.[34]

In many cases, the obligation of the peasantry to work without wages for the government prevented them from engaging in the few paid porterage jobs that were available. Kirkpatrick[35] has noted that on the Hitaura-Kathmandu route:

> The merchants are liable to be delayed more or less in their journey by the want of porters; and I was sorry to observe, that they appeared but too much exposed to it from the loose or arbitrary form of the government; as no ceremony was used in depriving them, for our accommodation, both at Hettowra and Goolpussra, of the carriers with which they had provided themselves.

Nor was this an isolated case. Kirkpatrick adds:[36]

> The evil would have scarcely merited notice, had it been limited to the particular case in question; but I am afraid the instances of it occur too often, when any of the principal men of the country happen to travel (especially on public business) in the route of the merchants.

Rana Policies

The Rana government initiated two measures to check such abuses and regulate the use of *jhara* labour: a blanket ban on such labour, and legislation aimed at prohibiting the use of unpaid labour by private individuals. The ban on the *jhara* system was imposed in 1847 on the ground that the system had subjected the people to great hardships. At the same time, the government reserved the right to employ people without wages "whenever it may be in need of such services." The ban was accordingly withdrawn early in 1854 when preparations were started for a war with Tibet.[37] It was never restored, and available evidence shows that unpaid labour obligations remained a normal feature of Nepal's agrarian life throughout the nineteenth century.

In 1854, legislation was enacted for enforcement all over the country on the subject of forced labour and wage-rates. Government officials and local functionaries were prohibited from exacting unpaid labour, except for meeting customary governmental requirements. They were required to pay wages at the rate of four *annas* daily, and a fine of the same amount, if they were proved to have forced people to work without wages in contravention of this ban. *Jagirdars* were required to pay wages at the rate of ten *paisa* daily if they exacted porterage services from the inhabitants of homesteads and villages assigned to them. However, owners of lands of all tenure categories were permitted to conclude agreements with their tenants stipulating the supply of unpaid labour for agricultural work or porterage.[38] In 1888, the statutory daily wage rate of four annas was abolished, and employers were required to pay only "reasonable wages fixed through mutual consent."[39] Both these measures, therefore, failed to introduce any real change in the condition of the peasantry.

Moreover, the ban on *jhara* labour, although short-lived, had no effect on the exaction of labour-tax under the *rakam* system. In fact, the latter half of the nineteenth century witnessed a considerable expansion in the scope of the *rakam* system. There were mainly two reasons for such expansion. In the first place, the growing centralization of the administration led to an increase in the volume of official correspondence, and hence to a more intensive exploitation of the *hulak* system in the hill region.[40] Secondly, there was an increasing trend toward the exploitation of the *rakam* system to meet the personal needs of members of the new political elite, particularly in Kathmandu Valley and the peripheral areas. Many peasant households in these areas were brought within the scope of the *rakam* system for the first time when any member of the Rana family set up an independent household.[41] There were also many cases in which *rakam* services were diverted from military purposes to meet the personal needs of members of the Rana family. Several *rakam* workers of Thankot in Kathmandu, for instance, had been employed in transporting military supplies during the 1855-56 Nepal-Tibet war. Subsequently, they were employed as porters to transport timber for the construction of palaces for the Ranas.[42]

When *rakam* services actually became redundant, the Rana government commuted them into cash levies in the same manner as its predecessors had commuted labour-tax obligations under the *jhara* system. The services of stone-workers employed to cut grinding stones for use in gunpowder factories were thus commuted in 1888, when machinery was introduced for the production of gunpowder.[43] Such commutation inevitably led to an increase in the burden of monetary taxation and thereby depressed the economic condition of the peasant, particularly when there was no alternative and gainful outlet for the use of his labour-power.

Several important developments occurred in the field of *rakam* labour-tax policy after the commencement of Rana rule. In 1854, legislation was enacted codifying the tenurial facilities and privileges of *rakam* workers. Landlords were forbidden to increase rents or other payments due from *rakam* workers, or to evict them and resume the lands for personal cultivation.[44] If *rakam* workers defaulted in the payment of rents, the headman was held responsible for evicting the defaulter and appointing another person to replace him, the obvious intention being to ensure that *rakam* services were not dislocated.[45] During 1854-55, a massive programme was started in Kathmandu Valley and the peripheral areas to redistribute rice-lands held by *rakam* workers in order to ensure that people who did the same work received land-allotments of equal size.[46] An attempt was thereby made to correlate the size of *rakam* land allotments with the quantum of obligations for *rakams* of each category. Thanks to the *rakam* land redistribution programme, *rakam* workers with relatively small holdings of rice lands benefitted at the cost of their more affluent neighbours.

Incidence of Labour-Taxation

The previous sections contained a brief survey of the different categories of labour-taxation under the *jhara* and *rakam* systems, with special reference to measures intended to provide a *quid pro quo* for *rakam* services. There was no *quid pro quo* in respect to labour taxation under the *jhara* system, hence the question determining the incidence of such taxation was fairly simple. The statutory wage-rate for an unskilled labourer during the nineteenth century was four *annas* per day. A peasant whose services were impressed thus indirectly paid a tax of that amount for each day of work. The burden was naturally higher if his *jhara* labour-tax obligations made it necessary for him to travel long distances from his residence, or if he had to provide tools and implements himself.

The incidence of labour taxation under the *rakam* system may also be assessed on the basis of the same formula. As mentioned previously, *Kagate-hulak* was the most extensively used *rakam* service in the hill regions. During the 1850s, 3,916 households had been enrolled under that system between Kathmandu and the western frontier in Doti, and 1,063 households in the east to Dhankuta, making a total of 4,979 *hulaki* households. Each household supplied one porter for the transportation of official mail for 96 days in the year.[47] The government thus exacted 477, 984 man-days of unpaid labour services under the *Kagate-hulaki* system. At the statutory wage-rate of four *annas* per day, this makes a total amount of Rs 119,496. The amount of homestead tax exemptions must, of course, be subtracted from that figure. Such exemptions normally amounted to less than a rupee per household.[48] Even assuming that they amounted to a rupee, the relief obtained thereby by *Kagate-hulaki* porters amounted to only Rs 4,979. The net incidence, consequently, amounted to Rs 114,517. In other words, the government of Nepal, during the mid-nineteenth century, exacted porterage services worth that amount from peasants in the eastern and western hill areas through the *Kagate-hulak* rakam alone.

There were several other factors besides homestead tax concessions which mitigated the incidence of labour taxation on the peasant, but it is difficult to measure the relief in monetary terms. As mentioned previously, *rakam* workers usually enjoyed exemption from labour taxation under the *jhara* system, which undoubtedly was more onerous and unpredictable. Moreover, *rakam* workers enjoyed a *quid pro quo* in the form of tenurial security as well as allotments of rice lands on a preferential basis. These facilities and concessions appear to have been tangible enough to induce peasants occasionally to offer to work under the *rakam* system on their own initiative,[49] or even to have their *rakam* obligations discharged through hired labourers, bondsmen or slaves.[50] Nevertheless, these indirect benefits cannot conceal the fundamental inequitability of the labour-tax system in nineteenth-century Nepal. The system did not provide an adequate *quid pro quo* to the peasant in consideration of the labour services that he was compelled to provide to the government, or to the landowning elites. In other words, the *jhara* and *rakam* systems denied the peasant an economic return for his labour services, prevented him from making use of his labour power for economic gain, and compounded the burden of taxation in money or commodities. Labour-taxation under these systems was, therefore, one more way in which the government, or the landowning elite, absorbed the economic surplus generated by the Nepali peasant during the nineteenth century.

NOTES

[1]William Kirkpatrick, *An Account of the Kingdom of Nepaul* (reprint of 1811 ed.), New Delhi, Manjusri Publishing House, 1969, p. 39.

[2]Cf. "Royal Order to Kaji Jaspau Thapa Regarding Unpaid-Labour Obligations on his Jagir Lands," Magh Sudi 8, 1887 (January 1831), *RRC*, Vol. 44, pp. 197-98.

[3]Cf. "Order Regarding Exaction of Unpaid Labour from Inhabitants of Salyan Rajya," Jestha Sudi 8, 1953 (June 1896), in Yogi Naraharinath (ed.), *Sandhipatra Sangraha* (A collection of treaties and documents), Dang, the editor, 2022 (1966), pp. 436-37.

[4]"Complaint of Tenants Cultivating Birta Lands in Ghoryagaun, Pyuthan," Shrawan Badi 11, 1908 (July 1851), *RRC*, Vol. 66, pp. 509-13.

[5]Kirkpatrick, *op. cit.*, pp. 101-2; Francis Hamilton, *An Account of the Kingdom of Nepaul*, (reprint of 1819 ed.) New Delhi, Manjusri Publishing House, 1971, p. 220.

[6]Naraharinath, *op. cit.*, p. 562.

[7]"Order Regarding Thek-Thiti Arrangements in Chinga Dara, Dullu-Dailekh," Marga Badi 5, 1938 (November 1881), *RRC*, Vol. 58, pp. 129-40.

[8]*Ibid.*

[9]"Revenue Regulations for the Eastern Tarai Region," Marga Badi 6, 1918 (November 1861), *ibid.*, Sec. 68, Vol. 10, pp. 50-51; "Revenue Regulations for the Naya Muluk Regions," Marga Badi 6, 1918 (November 1861), *ibid.*, Sec. 71, Vol. 47, pp. 470-71.

[10]Mahesh C. Regmi, *Land Tenure and Taxation in Nepal*, Berkeley, University of California Press, 1963-68, Vol. 3, pp. 54-56.

[11]In some Himalayan areas, such as Manang, the *Jhara* system appears to have been unknown before the Gorkhali conquests. "Commutation of Jhara Labour in Manang," Chaitra Sudi 3, 1854 (March 1798), *RRC*, Vol. 23, p. 305.

[12]"Abolition of Hulak, Beth, and Begar Systems in Morang," Bhadra Badi 5, 1847 (August 1790), *ibid.*, Vol. 19, p. 462; "Abolition of Jhara, Beth, and Begar Systems in Saptari," Jestha Badi 4, 1880 (May 1823), *ibid.*, Vol. 43, pp. 438-39. The ban on the exaction of unpaid labour in the eastern Tarai region,

except when it had been commuted into a cash levy, was reconfirmed by the Rana government in 1849. "List of Abolished Taxes and Levies in the Eastern Tarai Region," 1906 (1849), *ibid.*, Vol. 36, pp. 3-4.

[13]"Order Regarding Clearing of Nepal-India Boundary in the Butaul-Naya Muluk Region," Poush Badi 4, 1942 (December 1885), *ibid.*, Vol. 54, pp. 153-60; "Construction of Channel for Water Supply in Nepalgunj," Baisakh Badi 5, 1953 (April 1896), *ibid.*, Vol. 61, pp. 21-56.

[14]"Order Regarding Payment of Wages to Peons Employed to Transport Cash in Bara District," Shrawan Sudi 5, 1857 (July 1830), *ibid.*, Vol. 44, p. 54; "Administrative Regulations for the Western Tarai Region," Chaitra Badi 3, 1915 (March 1859), *ibid.*, Sec. 29, Vol. 47, p. 616.

[15]Regmi, *Land Tenure and Taxation, op. cit.*, Vol. 3, pp. 58-60; Mahesh C. Regmi, *A Study in Nepali Economic History, 1768-1846*, New Delhi, Manjusri Publishing House, 1971, pp. 103-9.

[16]Regmi, *Nepali Economic History, op. cit.*, pp. 116 and 187.

[17]"Thek-Thiti Arrangements in Mugu Village, Jumla District," Marga Badi 8, 1937 (November 1880), *RRC*, Vol. 58, pp. 61-75.

[18]Regmi, *Nepali Economic History, op. cit.*, pp. 104-5.

[19]Orfeur Cavanagh, *Rough Notes on the State of Nepal*, Calcutta, W. Palmer, 1851, pp. 15-17.

[20]Richard Temple, *Journals Kept in Hyderabad, Kashmir, Sikkim, and Nepal*, London, W.H. Allen, 1887, Vol. 2, p. 256.

[21]"Complaint of Inhabitants of Pyuthan Regarding Fiscal and Labour Obligations," Falgun Sudi 14, 1945 (March 1889), *RRC*, Vol. 62, pp. 39-53.

[22]"Fiscal and Labour Obligations of Inhabitants of Panchsayakhola in Nuwakot," Bhadra Sudi 8, 1912 (September 1855), *Regmi Research Series*, Year 7, No. 4, April 1, 1975, p. 78; and Poush Badi 10, 1935 (December 1878), *RRC*, Vol. 67, pp. 414-21.

[23]"Order Regarding Checkpost at Tilpung Bhanjyang," Poush Badi 1, 1937 (December 1880), *ibid.*, Vol. 55, pp. 727-31.

[24]Regmi, *Land Tenure and Taxation, op. cit.*, Vol. 3, p. 55.

[25]"Register of Kagate-Hulaki Households in Nepal," 1909 (1852), in *RRC*.

[26]"Order Regarding Enrollment of Hulaki Households in Majhkhand,"Jestha Badi 6, 1909 (May 1852), in Yogi Naraharinath, *op. cit.*, p. 385. Land survey regulations promulgated for the eastern hill region in 1895 similarly directed that "untouchable castes shall not be enrolled as *hulaki* porters." Register of Kagate Hulaki Rakam Households in East No. 1, 1952 (1895), *RRC*, Vol. 11, p. 351.

[27]Regmi, *Nepali Economic History, op. cit.*, p. 188.

[28]"Hulak Regulations for Areas Between Dharmathali and the Jamuna River," Ashadh Sudi 4, 1866 (June 1809), *ibid.*, Sec. 14, Vol. 6, pp. 1016-17.

[29]"Hulak Regulations for Areas Between the Bagmati River and Vijayapur," Baisakh Sudi 13, 1871 (May 1814), *ibid.*, Sec. 10, Vol. 41, pp. 460-61. This does not mean, however, that each *Hulak* household in these regions possessed a rice-land allotment. During the 1850s, 1,448 among the 3,915 *Kagate Hulaki* households in the Western hills possessed only homestead lands, whereas in the eastern hills the number was only 113 in a total of 928 households. In the

western hills, 2,467 households were allotted 198,056 *muris* of rice lands, making an average of approximately eighty *muris* of rice land for each *Kagate Hulaki* household. In the eastern hills, 928 households were allotted 62,775 *muris*, making an average of approximately 68 *muris*. Each allotment ranged in size from 7 *muris* to 160 *muris*, obviously because of differences in the size of the family and the availability of rice lands for allotment. "Register of Kagate Households in Nepal," 1852. (See n. 25 above).

³⁰"Order Regarding Land Allotments to Rakam Workers in Kathmandu Valley," Falgun Badi 12, 1905 (February 1849), *RRC*, Vol. 62, pp. 306-8.

³¹Regmi, *Land Tenure and Taxation, op. cit.*, Vol. 3, p. 72.

³²"Arrangements Regarding Hulaki Households in Dailekh," Bhadra Badi 6, 1886 (August 1839), in Chittaranjan Nepali, *Janaral Bhimsen Thapa ra Tatkalin Nepal* (General Bhimsen Thapa and Contemporary Nepal), Kathmandu, Nepal Sanskiritik Sangh, 2013 (1956), pp. 328-29. "Register of Kagate-Hulaki Households in Jumla," Falgun Sudi 10, 1949 (March 1893), *RRC*, Vol. 11, pp. 349-80.

³³Regmi, *Nepali Economic History, op. cit.*, p. 11.

³⁴*Loc. cit.*

³⁵Kirpatrick, *op. cit.*, p. 37.

³⁶*Loc. cit.*

³⁷"Imposition of Rasad Obligations During Nepal-Tibet War," Poush Badi 10, 1911 (December 1854), *RRC*, Vol. 56, pp. 258-65.

³⁸Ministry of Law and Justice, "Jhara Khetala" (On Unpaid Labour and Wages), *Shri 5 surendra Bikram Shahdevaka Shasankalama Baneko Muluki Ain* (Legal code enacted during the reign of King Surendra Bikram Shah Dev), Kathmandu, the Ministry, 2022 (1965), p. 84.

³⁹The amendment has been quoted in full in "Order Regarding Construction of Irrigation Channel in Kaski," Poush Sudi 9, 1946 (January 1889), *RRC*, Vol. 50, p. 279.

⁴⁰Regmi, *Land Tenure and Taxation, op. cit.*, Vol. 3, p. 63. The Nepal-Tibet war also led to a temporary spurt in the volume of both *Thaple* and *Kagate-Hulaki* obligations. Cf. "Order Regarding Hulak Services in Panchasayakhola," Falgun Sudi 14, 1911 (March 1853), *ibid.*, Vol. 33, pp. 219-20.

⁴¹Cf. "Order Regarding Byang Rakam Services in the Gardens of Commander-in-Chief Ranoddip Singh's Palace at Narayanhiti," Jestha Badi 5, 1923 (May 1866), *ibid.*, Vol. 55, pp. 142-47.

⁴²Regmi, *Land Tenure and Taxation, op. cit.*, Vol. 3, p. 62.

⁴³*Ibid.*, p. 61.

⁴⁴Ministry of Law and Justice, "Jagga Jaminko," (On Land), in *Shri 5 Surendra . . . Muluki Ain, op. cit.*, Sec. 12, pp. 21-22.

⁴⁵Ministry of Law and Justice, "Bali Na Tirnya Mohiko" (On default in the payment of rents), *ibid.*, Sec. 6, p. 47.

⁴⁶Regmi, *Land Tenure and Taxation, op. cit.*, Vol, 3, pp. 71-74.

⁴⁷"Register of Kagate-Hulaki Households in Nepal," 1909 (1852). (See n. 25 above).

⁴⁸In 1856, for instance, when a new mail-route was opened through Udayapur, 45 local households were enrolled as mail-carrier and granted a total home-

stead-tax remission of Rs 3. Each household thus obtained a remission of 10 *annas* on an average. "Order Regarding Hulak Services in Udayapur," Kartik Sudi, 10, 1927 (November 1870), *RRC*, Vol. 55, pp. 475-77.

[49]"Order Regarding Hulak Services in Devpur and Elsewhere," Marga Badi 11, 1949 (November 1883), *ibid.*, Vol. 11, pp. 385-92.

[50]A reference to the practice of hiring labourers to perform *rakam* service is contained in "Hulak Regulations for the Bishnumati-Bheri Region," Baisakh Badi 7, 1885 (April 1828), *ibid.*, Secs. 2-3, Vol. 27, pp. 81-82. Identical regulations were promulgated on the same date for the Bheri-Mahakali region also, *ibid.*, Vol. 27, pp. 83-84. See also "Hulak Regulations for the Mechi-Vijayapur Region," Kartik Sudi 9, 1889 (October 1832), *ibid.*, Sec. 5, Vol. 27, p. 265; "Sadar Dafdarkhana's Order to the Kagate-Hulakis of Luintel-Bhanjyang," Magh Sudi 10, 1918 (January 1862), *ibid.*, Vol. 62, pp. 391-92, and "Complaint of Kalu Lama, a Bondsman," Falgun Badi 7, 1918 (February 1862), *ibid.*, Vol. 47, p. 743.

7

The Agrarian Community

For the various groups of the landowning elite who were entitled to
a share in the peasant's produce—*rajas, birta*-owners and *jagirdars*
—land was a dependable source of income, a secure form of property, and a symbol of social status. These benefits, however,
accrued only if the land was actually cultivated. Therein lay the
importance of the peasant, who performed the essential tasks of
ploughing the fields, sowing seeds, and harvesting crops. The peasantry, however, did not constitute a homogeneous group in the society.
A bi-polar view of Nepal's agrarian society during the nineteenth
century, with the landowning elite at one end of the spectrum and
the tiller of the soil on the other, would, therefore, bear no correspondence with the reality. In this chapter, we shall try to identify
the various groups in the local peasant community over which the
authority of the landowning elite was superimposed under the
rajya, birta and *jagir* systems. We shall also discuss the measures
taken by the Rana government to introduce some measure of tenurial uniformity in all parts of the country, solve the growing problem of eviction of tenants, and grant property rights in land to the
peasant. For the purpose of the present discussion, we shall assume
that land becomes a form of private property if it can be bought
and sold, or pledged as collateral for a loan.

Landholders and Peasants

Economic differences among different groups in the peasant community are the end-product of various social, economic, historical

and other factors which it would be out of place to discuss here. These differences are usually measured on the basis of such criteria as the size of holdings. Unfortunately, no reliable information in this regard is available for nineteenth-century Nepal. Accordingly, even though a discussion of the legal status of different peasant groups is a poor substitute for an analysis of the economic and other factors that determine his actual economic condition, the present state of our knowledge leaves us no choice of an alternative methodology. In the context of the present study, it may be sufficient to point out that individuals and groups who claimed ownership and, consequently, rent-receiving rights in the land, had perforce to come to terms with the essential role of the peasant as a primary producer. They had to concede him certain basic rights in order to ensure that he had adequate incentive to stay on and cultivate the land. These included the right of each local household to a homesite and an allotment of rice land, the right to occupy these lands and homesites so long as the peasant paid the customary rents and taxes, and the right to leave them to his successors through inheritance and partition. In those areas where the peasant had been able to strengthen his rights and privileges in the land by law and/or custom in such a manner that he enjoyed unchallenged possession of his holding, subject only to the payment of rents and taxes, he was often able to transfer his holding to another person on payment of a sum of money. Where the peasant held the lands that he tilled only on short leases from the government, or from the landowning elite, his right to his holding was not secure enough to attract monetary investments from prospective buyers.

In the present study, we shall classify peasant groups on the basis of the extent of these rights that they actually enjoyed. If purely local variations are ignored, the following categories stand out prominently in the agrarian community: peasants who cultivated the land at the will of the landowning elite, without any protection against arbitrary eviction, peasants who enjoyed some measure of security against arbitrary eviction, but were denied the right to sell or mortgage the land, and peasants who enjoyed statutory security of tenure, as well as the right to sell or mortgage the land. The term landlord will here be used to denote individual members of the landowning elite groups of *rajas*, *birta*-owners, and *jagirdars*. Actual cultivators will be described as peasants, while the term landholder will be used for individuals belonging to intermediary groups between the landlord and the peasant whose rights had been made secure by law or custom.

Resident Birta-owners and Kipat-owners

From the viewpoint of legal status and economic opportunities, it seems beyond doubt that small *birta*-owners, who lived in the villages where their *birta* lands were situated, formed the topmost layer of the local agrarian community. They were freeholders who were under no obligation to share their produce with a landlord. Even when a resident *birta*-owner and a *jagir* tenant cultivated holdings of the same size, therefore, the former was much better off. Moreover, *birta*-owners traditionally had legal authority to force their tenants to work without wages. They were thereby able to cultivate bigger farms at costs of production which were lower than those of a peasant cultivating *jagir* and other taxable lands. These social and economic privileges of resident *birta*-owners were not available to an ordinary peasant in the same village who cultivated his small plot of *jagir* land, paid rents and a multitude of homestead and other taxes, and in addition, provided unpaid labour services to the state and his landlord.

This does not mean, however, that resident *birta*-owners as a class necessarily occupied an economic status superior to the mass of the peasantry who cultivated *jagir* and other taxable lands in the capacity of tenants. The topmost layers, no doubt, consisted for the most part of resident *birta*-owners, but some of them must also have occupied holdings that were comparatively small by local standards. The reason is that a market in *birta* land had come into existence long before *jagir* and other *raikar* lands could be sold and purchased. The progressive fragmentation of old *birta* holdings must similarly have depleted their size to a considerable extent. We can only conclude, therefore, that because of the burden of taxation in money, commodities and labour, an individual peasant cultivating *jagir* and other lands of taxable categories would tend to be much poorer than an individual *birta*-owner-cultivator with a holding of the same size.

The position of *kipat*-owners, particularly those in the far-eastern hill region of Pallokirat, resembled that of *birta*-owners in many respects. *Kipat*, as already defined in Chapter 3, was a communal form of land tenure under which the concerned ethnic group, represented by the headman, rather than the government, controlled the allotment of land. Members of such ethnic groups owned land under the *kipat* system by virtue of their membership of the group. Earlier studies on the *kipat* system have mainly stressed its communal aspect, and described its customary characteristics vis-a-vis the statutory forms of tenure such as *raikar* and *birta*.[1] From the viewpoint of the local agrarian community, however, it appears necessary to lay equal emphasis on the intra-communal aspects of *kipat* tenure, that is to say, the relationship between the headman, in whose name royal orders confirming the customary occupation of lands under *kipat* tenure were usually issued during the eighteenth and nineteenth centuries, and the ordinary members of the *kipat*-owning community who subsisted on allotments of *kipat* lands made by the headman. The beneficiaries of such royal orders were known as *zamindars*,[2] or landlords. They often had their lands cultivated through sharecroppers, who paid "one half of the produce for rent."[3] During the early nineteenth century, *kipat* lands appear to have been freely sold[4] and mortgaged.[5]

Peasants and Landholders in Raikar Land

Resident *birta*-owners and *kipat*-owners were, however, islands of autonomy in an agrarian society where the superimposition of the landowning elite's authority was a virtually universal phenomenon. The character of the local peasant and landholding groups who were subjected to that authority was different in different regions of the Kingdom: the Tarai region, the Baisi region, and the central hill region.

(1) The Tarai Region

In the Tarai region, land during the nineteenth century was normally in the possession of two categories of landholders, *zamindars* and *chuni* ryots. The rights of *zamindars* "extended over land occupied by a number of persons,"[6] whereas a *chuni* peasant was only "an ordinary occupant or holder of land" whose name was listed in the official tax-assessment register.[7] The *zamindars* of the Tarai region during the nineteenth century comprised, as in India, "a rural class other than, and standing above, the peasantry."[8] No information is available as to how the *zamindars* of Nepal acquired that status in the first place. It is possible that, as in the adjoining areas of northern India, they belonged to castes or communities who had settled a new tract of territory, or occupied an inhabited tract after driving out or establishing their domination over the existing settlers.[9] In any case, their customary rights and status appear to have been sufficiently entrenched to be recognized by successive governments.

The *zamindari* system appears to have been more or less of a similar nature in both Nepal and the adjoining areas of northern India until the last quarter of the eighteenth century. It started evolving on divergent lines only after the permanent settlement was introduced in Bihar and Bengal in 1793. The settlement permitted *zamindars* "to transfer to whomsoever they may think proper, by sale, gift, or otherwise, their proprietary rights in the whole or any part of their respective estates without applying to government for its sanction to the transfer."[10] In Nepal, on the other hand, the government, instead of recognizing the proprietary rights of *zamindars*, preferred to deal directly with the cultivator.[11] Consequently, whereas the *zamindars* of India became proprietors of the land, the landholding rights of their counterparts in Nepal were virtually ignored.

The term *chuni* denoted peasants who obtained allotments of waste lands from the local authorities and so lay beyond the jurisdiction of *zamindars*. They appear to have been less secure in their tenure than *zamindars*, because local authorities often evicted them after they had reclaimed waste lands in order that the lands might be reallotted to other persons on more favourable terms. Such malpractices had been banned in 1793, but the ban appears to have been largely ineffective.[12] That may have been the reason why Hamilton noted that "were property somewhat more secure, this (Tarai) territory is capable of yielding considerable resources."[13]

There is no evidence that either *zamindars* or *chuni* landlords were legally entitled to sell their lands. Because agrarian systems in the Tarai region of Nepal had much in common with those prevalent in the adjoining areas of India during the eighteenth century, it may be correct to presume that the nature of rights in the land enjoyed by both *zamindars* and *chuni* ryots were more or less similar on both sides of the border at that time. In India, the *zamindar* had "no power of alienating his estate, he could not raise money on it by mortgage, nor sell the whole or any part of it."[14] Similarly, "long occupancy had not certainly empowered the ryots to sell or mortgage their lands."[15] In any case, it is possible that the relative abundance of cultivable lands, and the facilities that were usually offered to prospective settlers, constituted more or less effective constraints in the development of a market in land in the Tarai region. But even though private property in land may have been absent in the Tarai region in the sense of the right to sell or mortgage it, the rights of both *zamindars* and *chuni* landlords were sufficiently secure in law and custom to permit a cleavage between landholding and actual use. Their lands were cultivated for the most part by a class of sharecroppers called *adhiyars*.[16]

(2) *The Baisi Region*

The division of the upper layers of the agrarian society into *zamindars*[17] and *chuni* landholders[18] was a normal feature of the Baisi region also. The area of lands controlled by them was often so large[19] that "some people have kept their lands uncultivated, whereas others complain that they do not have adequate lands for their subsistence."[20] The large size of holdings often deterred personal cultivation by the owner, hence tenancy appears to have been commonly practised.[21]

There, however, ended the similarity between the agrarian societies of the Tarai and Baisi regions. Whereas a market in land had remained undeveloped in the Tarai region, in the latter both sale and mortgage transactions had been customary[22] even before the Gorkhali conquest of that region during the late eighteenth century.[23]

(3) *The Central Hill Region*

The agrarian society of the central hill region presented a glaring contrast to that in the Tarai region and the Baisi region. There it consisted largely of peasants with small, self-cultivated rice-land allotments. Preference in such allotment was usually given to resident peasants, that is to say, peasants who lived in the village or area where the lands were situated. Customary law provided that no peasant should be deprived of his homestead and rice-land allotment so long as he remained in occupation and paid the stipulated rents and other dues. However, the tenurial rights provided by customary law were confined to the actual use of the land. Possession of land independently of actual use was seldom possible, nor had a market in land developed. Indeed, during the first decade of the nineteenth century, regulations were enforced in the central hill region making transactions in land a punishable offence.[24]

The tenuous nature of the peasant's rights in land in the central hill region is highlighted by a new system of rice-land allotment, known as *raibandi*, that the government introduced during the 1830s on *jagir* lands assigned to the army.[25] Under the *raibandi* system, *jagir* lands of that category were reallotted among the existing tenants in such a manner that each cultivator received an allotment of rice-land that he could cultivate through the labour of his family.[26]

What were the circumstances that led the government to introduce the *raibandi* system? The growing requirements of rice-lands for assignment under the *jagir* system necessitated the reclamation of large areas of marginal lands high up on the hill sides and in river banks through the construction of irrigation channels and embankments. In most cases, these channels and embankments were earthen constructions which had to be repaired each year during the winter season. Lands were allotted on a yearly basis to the local peasants on condition that they contributed unpaid labour for these repairs.[27] Notwithstanding these precautions, the newly-reclaimed lands continually faced the threat of erosion. Moreover, *jagir* holdings often comprised lands which could be used to grow rice but had not actually been so used because of the high costs of irrigation or risks of damage. The government, therefore, needed a system that would compel the peasant to bear the burden of *jagir* rents on waste lands and the risks of loss through damage to the land as a result of erosion. The *raibandi* system helped to meet these needs.

Under the *raibandi* system, therefore, available land resources were reallocated in the local community in such a manner that every adult inhabitant became a tax-payer. The correlationship that the system established between the size of the rice-land holding and the size of the peasant's family minimized the risk of loss resulting from any individual peasant's default in the payment of rents and taxes. After the introduction of the Rana regime, the ambit of the *raibandi* system was widened to include all categories of *raikar* and *jagir* lands, as well as *guthi* and *kipat* lands.

Notwithstanding the advantages that the *raibandi* system may have brought to the government, it undermined the permanence of the peasant's tenure. It weakened the force of the customary law according to which a peasant had the right to occupy his holding so long as he paid the stipulated rents and other customary dues,

A family whose rice landholding exceeded the *per capita* share in the village was obliged under the *raibandi* system to relinquish the excess area without any compensation.[28] In other words, a peasant family which had an inadequate holding could get a subsistence holding through the reallotment of the bigger holdings of its more affluent neighbours. Under the *raibandi* system, therefore, a peasant's rights over the lands cultivated by him became neither permanent nor inheritable and subdivisible.

Impermanence of tenure should not, however, be confused with insecurity. The *raibandi* system only meant a redistribution of fields; in no case did it deprive any peasant of a subsistence holding. In fact, by conceding the right of each member of the local peasant community to such a holding, the system may actually have provided the community with a large measure of security and stability.

The agrarian structure that emerged through the *raibandi* system was necessarily characterized by small cultivators. Although holdings might differ in size because of differences in the size of the family, the *per capita* cultivated area was more or less equal. There is evidence, however, that such redistribution did not create an egalitarian agrarian society. There were several reasons for this situation. Under the *raibandi* system, lands were redistributed only among the local cultivating households. Households that possessed no rice lands apparently found no place in the list of beneficiaries under the *raibandi* system. Moreover, *raibandi* allotments were made on rice-lands only. No attempt was made to regulate the size of homesteads. Nor was this all. A peasant who obtained a *raibandi* allotment could also cultivate non-*raikar* lands concurrently, if he could get any, in the capacity of a tenant. Finally, rice-lands were often allotted directly by the central government from Kathmandu. Such holdings were excluded from the purview of redistribution under the *raibandi* system.

To sum up, landholding groups in the local agrarian community who could retain their rights in the land without cultivating it themselves, and, occasionally, even alienate it through sale, mortgage, or allotment, were confined for the most part to the Tarai region, the Baisi region, and the far-eastern hill region. The agrarian society of the central hill region, in contrast, consisted for the most part of small cultivators who held their lands on short leases and usually forfeited their rights if they did not cultivate the land themselves. Such impermanence of tenure prevented the emergence of tenancy and of a market in land in that region.

The Role of Landholding Groups

During the early nineteenth century, official policy was generally aimed at undermining the control that *zamindars, kipat*-owners, and other landholding groups in different parts of the country exercised over the agricultural lands under their jurisdiction. This policy was dictated by several considerations of expediency. The Gorkhali rulers aimed at implementing as far as possible the theory that the state is the owner of all lands and other natural resources in its territories. The reason for such a policy was that land in early nineteenth-century Nepal was important to the government not because it yielded tax revenue but because it could support an army and a bureaucracy through assignments under the *jagir* system. Such assignments resulted in the superimposition of the *jagirdar's* authority over the local agrarian community. A multilayered structure of landholding rights in the local agrarian community obstructed the unrestricted exercise of the *jagirdar's* authority to evict recalcitrant cultivators or raise rents. Consequently, one of the main objectives of Gorkhali agrarian policy was to bypass the landholding groups and promote a direct relationship between the landowning elite and the peasant. Where the rights of the landholding groups had been too strongly entrenched to make possible such a direct relationship with the peasant, they attempted to reduce the area of land controlled by such groups.

In the Tarai region, for instance, the government introduced a land tenure system which, in India, has been described as *ryotwari.* The system ignored intermediary landholding groups and assessed taxes directly on the cultivating peasant. It thus ignored the status of *zamindars* as "a rural class other than, and standing above, the peasantry." A similar policy was followed in the Baisi region as well. The redemption of lands mortgaged before 1806 by landholding groups was prohibited. Lands belonging to landholders who did not cultivate them personally were taken away from them, and granted to the tenants. The expropriated landowners were forbidden to collect rents and other payments, or to exact unpaid labour, from their erstwhile tenants. Actual cultivators were thus enabled to hold lands in their own names, whereas the owners were expropriated without compensation. Often such landowners were given the choice of either paying taxes on their uncultivated lands, or letting the local administration reallot these lands to persons who were willing to cultivate them.[29] The objective of these measures was to increase the number of households and check tenurial inequities and concentration of landholding rights. With the same objective, village functionaries were instructed not to wipe out existing homesteads "by purchasing lands and people."[30] There is evidence, however, that these reforms were not effectively enforced, with the result that the dispossessed landowners were often able to continue collecting rents from their erstwhile tenants. The latter had then to bear the burden of both these payments and the newly-imposed state taxes.[31]

In the case of *kipat*, which was customarily tax-exempt, conversion into *raikar* made the lands subject to taxation, and so increased the area available for grant or assignment as *birta* or *jagir*. The official policy was, therefore, to bring alienated *kipat* lands within the ambit of the *raikar* tax system. During the last years of the eighteenth century, the government also imposed ceilings on *kipat* holings in the eastern hill areas. There were also other considerations of a political nature behind the government's efforts to convert *kipat* lands into *raikar* which are not relevant to the present discussion.[32]

Efforts to establish a direct relationship between the landowning elite and the peasant had, nevertheless, only a palliative effect; they did not seek to introduce structural changes in the local agrarian community. Consequently, the system of intermediary landholding remained unaffected. It is, therefore, necessary to discuss the position that the peasant occupied as the lowest stratum of the three-tiered hierarchy of landed interests, with the landowning elite or the government occupying the topmost position.

Landholding groups who occupied an intermediary position between the landowning elite or the government on the one hand, and the peasantry on the other, also claimed a share in the surplus produce of the peasantry. However, this did not necessarily mean that the peasant who worked under a *zamindar* in the eastern Tarai region bore a cumulated burden of taxes and rents that was heavier than the burden borne by his counterpart in the central hill region. Irrespective of the form of tenure, the peasant who actually cultivated the land paid at least one half of the crop as rent. Intermediary landholding groups emerged in the Tarai and the Baisi region, but not in the central hill region, primarily because the incidence of land taxation was lower in the former; and they could subsist on the difference between the rate of rent paid by the cultivator and the tax they themselves were under obligation to pay to the government. In either case, the peasant who actually cultivated the land paid a rent which usually amounted to at least half of his produce.

The fact that the landholding rights of intermediary groups had been established by custom, and subsequently granted statutory endorsement, did not, therefore, necessarily mean that the ordinary peasant in the Tarai and the Baisi region occupied a status higher than that of his counterpart in the central hill region. As previously noted, these rights had in many cases resulted in a divorce between ownership and actual use, with the consequence that they were possessed by the non-working owner, rather than by the actual cultivator. In most of the country, therefore, the rights of the peasant in the lands he tilled were limited to usufruct.

Position of the Peasant

The key element in this discussion of the hierarchy of agrarian groups in different regions of the kingdom during the nineteenth century is the position of the peasant at the lowest rung. Where the government dealt directly with the peasant, as in the central hill region, the peasant was treated no better than a tenant-at-will. On the other hand, where select groups had attained a privileged status in the local agrarian community, such as the *zamindars* of the Tarai and Baisi regions, the peasant was left at their mercy, and rarely became an object of concern for the government. The security of the peasant's tenure in these circumstances largely became a matter of custom, liable to be violated at any time when the local landholding groups could lay their hands on another prospective tenant on more favourable terms. The most that the peasant could expect to achieve in any part of the country during the early nineteenth century was the customary ban, often reinforced by administrative sanction, on arbitrary eviction so long as the stipulated dues and exactions were paid.

It is essential, however, to keep the problem of tenurial security in a proper perspective. Although no statistics are available, a study of Nepali source materials during the late eighteenth and early nineteenth centuries inevitably leads to the conclusion that the Kingdom faced a shortage of manpower in relation to available land resources. The problem of labour shortage was, in fact, serious enough to justify a liberal policy on immigration, particularly into the Tarai from the adjoining areas of northern India. It also fostered a competition between the government and the landowning elite for prospective tenants to cultivate their lands. Local administrators in the Tarai region, for instance, were repeatedly instructed to lure peasants from *birta* lands for the cultivation of taxable lands under the control of the government.[33] Demographic factors thus made the eviction of tenants a proposition of doubtful benefit to the landowning elite.

Emergence of Property Rights

By the middle of the nineteenth century, several developments occurred in the central hill region which had an adverse impact on the security of the peasant's tenure. The introduction of the *kut* system for the assessment of rents on rice-lands was the most important of these developments. So long as the actual crop was divided equally between the landlord and the peasant under the *adhiya* system, it mattered little to the landlord who actually cultivated the land. He had, indeed, little to gain by replacing an existing tenant by a new one. But the *kut* system put an end to this stability. Because rents were fixed in advance at specific rates under the *kut* system, a newcomer could stipulate a higher amount than the incumbent. The landlord had then a strong motive to evict his tenant and accept the higher offer. The government accordingly enacted legislation which gave landlords full authority to evict peasants who refused to accede to their demands for higher rents.[34] Peasants' customary rights in the land were thus sacrificed at the altar of the *jagirdar*'s greed.

To be sure, the customary law that peasants should not be evicted so long as they paid the stipulated rents and taxes was reiterated from time to time through administrative orders in different areas of the country during the early nineteenth century.[35] There is also evidence that the government occasionally realized the a_verse impact of tenurial insecurity on agricultural production and the economic condition of the peasant and made sporadic attempts to mitigate that impact.[36] Nevertheless, there is no evidence that any attempt was made during the early nineteenth century to deal with the problem of tenurial insecurity on a comprehensive basis.

Rana Policies

Measures were initiated only after the emergence of the Rana regime to provide tenurial security to peasants in all parts of the country. The Rana rulers aimed at preventing the growth of the *jagir*-owning bureaucracy from entrenching itself as a powerful and autonomous landed interest. They sought to achieve this objective by freeing the peasantry from excessive dependence on their *jagir-dar*-landlords.[37] As already noted in Chapter 4, the 1854 legal code specified the circumstances in which landlords were permitted to evict their tenants. It provided full security of tenure to peasants so long as they paid the stipulated rents and taxes. Simultaneously, records of rights were compiled, thereby providing the peasant with documentary evidence of title to the land he tilled. These developments set the stage for the emergence of property in land. If a peasant's holding was favourably located, or was of high fertility, or if he had invested labour and capital to make it productive, he was now able to sell it, or take a loan and give the land on mortgage. By the third quarter of nineteenth century, peasants cultivating *jagir* and other categories of *raikar* lands in all parts of the country had succeeded in acquiring property rights.[38] The improvement that the Rana land reform measures brought about in the status of the *raikar* peasantry as property owners thus appears to have been substantial.

At the same time, the inadequacies of these measures should not be overlooked. They provided protection only to the *raikar* peasantry. As noted in Chapter 5, the 1854 legal code granted full authority to *birta*-owners to evict their tenants in expectation of higher payments from others. A large segment of the peasantry, consisting of those who cultivated *birta* and other tax-free lands, were, consequently, left at the mercy of their landlords. Nor was this all. The tenurial security and property rights that the *raikar* peasant was able to obtain thanks to the Rana land reform measures made it possible for him to sell or mortgage his lands, particularly as growing population intensified the demand for land. Many peasants found it more profitable to expand their holdings and sublet the lands than to cultivate them personally.[39] They were thereby able to subsist on the differences between the rents they received and the taxes they paid to the government or to the *jagir-dar*-landlord. In the event of such subinfeudation, the tenurial security and property rights provided by the land reform measures benefitted not the actual cultivator, but the non-working intermediary.

We may now recapitulate the main lines along which the tenurial rights of peasants on *jagir* and other categories of *raikar* lands evolved under the impact of policies initiated by the Rana government. The chief feature of these policies was that the cultivator was guaranteed security of tenure, and that the compilation of records of rights made it possible for him to enjoy that security in actual practice. Property rights in land emerged as a consequence of this development. Such rights enabled the cultivator to sell or mortgage his holding, thereby virtually obliterating the long-standing differences in the tenurial status of the peasantry in the Baisi region, the Tarai region, and other parts of the country. Rana rule thus witnessed a definite trend toward the strengthening of the peasant's right to own land as property. However, that right was at times appropriated not by the actual cultivator but by an intermediary landholder or a moneylender.

NOTES

[1] Mahesh C. Regmi, *Land Tenure and Taxation in Nepal*, Berkeley, University of California Press, 1963-68, Vol. 3, pp. 82-95; and *Landownership in Nepal*, Berkeley, Los Angeles and London, University of California Press, 1976, pp. 87-103.

[2] "Order Regarding Land Reclamation in Tumlingtar," Shrawan Sudi 5, (July 1870), *RRC*, Vol. 55, p. 458.

[3] Francis Hamilton, *An Account of the Kingdom of Nepal* (reprint of 1819 ed.) New Delhi, Manjusri Publishing House, 1971, p. 149.

[4] "Order Regarding Transactions in Kipat Land," Jestha Badi 9, 1943 (May 1886), *RRC*, Vol. 51, pp. 682-90.

[5] "Judicial Regulations for the Eastern Hill Region," Marga Badi 9, 1866 (November 1809), *ibid.*, Sec. 12, Vol. 40, p. 127.

[6] Irfan Habib, *The Agrarian System of Mughal India*, Bombay, Asia Publishing House, 1963, p. 140.

[7] "Regulations for the Western Hill Region," Kartik Sudi 1, 1893 (October 1836), *RRC*, Sec. 9, Vol. 35, p. 66.

[8] Habib, *op. cit.*, p. 141.

[9] *Ibid.*, p. 160.

[10] B.H. Baden-Powell, *The Land Systems of British India* (reprint of 1892 ed.), Delhi, Oriental Publishers, n.d., Vol. 1, p. 410.

[11]Mahesh C. Regmi, *A Study in Nepali Economic History, 1768-1846*, New Delhi, Manjusri Publishing House, 1971, p. 92.

[12]*Loc. cit.*

[13]Hamilton, *op. cit.*, p. 64.

[14]Baden-Powell, *op. cit.*, Vol. 1, pp. 397 and 513.

[15]Ram Narayan Sinha, *Bihar Tenantry (1783-1833)*, Bombay, People's Publishing House, 1968, p. 81.

[16]Regmi, *Nepali Economic History*, p. 32.

[17]Cf. "Confirmation of Jimidari Holding Granted by King of Jumla," Bhadra Sudi 10, 1851 (September 1794), *RRC*, Vol. 5, p. 748.

[18]"Regulations for the Western Hill Region," 1836, sec. 9. (See n. 7 above).

[19]Such holdings were known as *rekh*. "Confirmation of Landownership Rights in Jumla," Ashadh Sudi 15, 1915 (June 1858); *RRC*, Vol. 66, pp. 313-16. See also "Confirmation of Rekh Lands of Barmas and Bhats in Jumla," Jestha Badi 10, 1901 (May 1844), in Yogi Naraharinath (ed.), *Itihas Prakash* (Light on History), Kathmandu, Itihas Prakash Mandal, 2012-13 (1955-56), Vol. 2, Bk 2, p. 232. For the text of a land grant of this category, made by a King of Dullu in A.D. 1588, see "Land Grant to Ranu Kanya and Kusum Padhya," Falgun 7, 1644 (February 18, 1588), in Yogi Naraharinath (ed.), *Sandhipatrasangraha* (A collection of treaties and documents), Dang, the editor, 2022 (1966), p. 379.

[20]"Order Regarding Cultivation of Lands in Doti," Ashadh Badi 4, 1872 (June 1815), *RRC*, Vol. 43, p. 13; "Complaint of Katuwals of Chilkhagaun, Jumla," Kartik Sudi 9, 1888 (October 1831), *ibid.*, Vol 38, p. 223.

[21]"Confirmation of Landownership Rights in Jumla," 1858 (See n. 19 above).

[22]"Deed of Land Sale in Ukhadigaun, Jumla," Kartik Sudi 14, 1895 (November 1838), in Naraharinath, *Itihas Prakash, op. cit.*, Vol. 2, Bk 2, pp. 177-78; "Regulations for Dullu and Dailekh," Aswin Sudi 4, 1879 (September 1812), *RRC*, Vol. 43, pp. 360-63.

[23]Regmi, *Nepali Economic History, op. cit.*, p. 30.

[24]*Ibid.*, p. 70.

[25]"Order Regarding Reallotment of Jagir Lands in Chisankhu," Marga Badi 9, 1887 (November 1830), *RRC*, Vol. 44, p. 149; "Complaints Regarding Raibandi Reallotment of Jagir Lands of Sabuj and Ser Battalions," Jestha Badi 2, 1905 (May 1848), *ibid.*, Vol. 62, pp. 690-94. According to the latter document, the *raibandi* system had been introduced on these lands in A.D. 1839.

[26]"Order Regarding Raibandi Land Allotments in Thansing," Magh Sudi 2, 1902 (January 1846), *ibid.*, Vol. 7, pp. 505-9; "Orders Regarding Land Allotments in Kaski and Other Areas," Magh Badi 11, 1894 (January 1838), *ibid.*, Vol. 27, pp. 499-501.

[27]Cf. "Orders Regarding Repair of Pardi Embankment in Pokhara," Poush Badi 30, 1890 (December 1833), *ibid.*, Vol. 27, p. 245 and Marga Badi 5, 1908 (November 1851), *ibid.*, Vol. 27, p. 245 and Marga Badi 5, 1908 (November 1951), *ibid.*, Vol. 66, pp. 549-52.

[28]"Raibandi Land Allotments in Dhor," Magh Sudi 2, 1921 (January 1865), *ibid.*, Vol. 21, pp. 613-17; "Order to Bichari Tikanidhi Lohani of Gorkha

Adalat," Baisakh Badi 4, 1924 (April 1867), *ibid.*, Vol. 32, pp. 266-73.

[29]This account is based on the following sources: "Order to the Rajawars of Doti," Ashadh Badi 4, 1872 (June 1815), *ibid.*, Vol. 42, p. 13; "Regulations for Jumla," Chaitra Badi 8, 1886 (March 1830), *ibid.*, Vol. 34, pp. 54-55. "Thek-Thiti Arrangements in Sinja, Jumla District," Chaitra Badi 7, 1894 (March 1838), *ibid.*, Vol. 35, pp. 590-60; and "Confirmation of Customs and Usages of Jumla," Chaitra Sudi 7, 1900)March 1844), *ibid.*, Vol. 34, pp. 609-19.

[30]"Thek-Thiti Arrangements in Sinja, Jumla District," 1838. (See n. 29 above).

[31]*Ibid.*, "Order Regarding Unauthorized Collection of Rents in Jumla," Bhadra Badi 2, 1885 (August 1828), *ibid.*, Vol. 43, pp. 119-21.

[32]Regmi, *Land Tenure and Taxation, op. cit.*, Vol. 3, pp. 123-25, and *Landownership in Nepal, op. cit.*, pp. 92-101.

[33]Regmi, *Nepali Economic History, op. cit.*, p. 146.

[34]*Ibid.*, p. 182; "Order Regarding Rice Land Allotments in the Chepe/Marsyangdi-Bheri Region," Poush Sudi 6, 1885 (January 1829), *RRC*, Vol. 43, pp. 161-62.

[35]Regmi, *Nepali Economic History, op. cit.*, p. 186.

[36]Cf. "Order Regarding Kut Rent Payments in Limi, Nuwakot District," Jestha Badi 12, 1892 (May 1835), *RRC*, Vol. 45, p. 97.,

[37]S.N. Eisenstadt, *The Political System of Empires*, New York, Free Press, 1963, p. 133.

[38]Regmi, *Landownership in Nepal, op. cit.*, pp. 174-77.

[39]Regmi, *Land Tenure and Taxation, op. cit.*, Vol. 3, pp. 56 and 76.

8

The Village Moneylender

Our study of Nepal's land tenure and taxation systems in the previous chapters showed how surplus agricultural production and the
peasants' labour power were absorbed by the government or by
the landowning elite, *rajas*, *birta*-owners, and *jagirdars*, as well as
by village headmen and other categories of functionaries employed
by them. The peasant was compelled to bear the burden of sustaining this hierarchy of landed interests because, under the existing
land tenure system, all land belonged either to the government or
to individuals designated by it under the *rajya*, *birta* and *jagir* systems. The authority enjoyed by *rajas*, *birta*-owners and *jagirdars*,
as well as village headmen and other functionaries who collected
revenue on their behalf, depended on their political power, and
bore no relationship with the services or benefits that they made
available to the peasant. The obligations that the peasant owed to
those landed interests were, consequently, a manifestation of political domination. We shall now discuss how yet another category
of agrarian interests, the moneylenders, supplied the credit needs
of the peasant, and in return, exploited his labour power, or appropriated a share in his production. The peasant's obligations to
the moneylender were, therefore, a manifestation of economic
domination.

The Problem of Agrarian Indebtedness

Agrarian indebtedness appears to have been a chronic and ubiquitous problem in nineteenth-century Nepal. Subsistence farming

at a low level of productivity, which was characteristic of Nepal's agriculture during this period, compelled the peasant to borrow from moneylenders when crops failed, or when death or illness in the family hindered normal cultivation. The adverse effect of agrarian indebtedness on agricultural production and the stability of the rural population was serious enough to invite governmental measures aimed at controlling rates of interest and scaling down the amount of loans. Even before the political unification of the Kingdom, King Ram Shah of Gorkha (1606-36) had prescribed the rate of interest on cash loans at 10 per cent every year, and on loans in kind at 25 per cent, subject to a maximum payment of twice the amount of the loan if payments had accumulated.[1] During the eighteenth and nineteenth centuries, these provisions were reconfirmed from time to time,[2] and additional regulations were promulgated prescribing that amounts collected as interest in excess of these rates should be deducted from the principal.[3]

However, agrarian reform measures which militate against the interests of elite groups such as moneylenders require considerable administrative effort and organization to be effective. There is no evidence that these regulations were backed by such effort and organization. The repeated promulgation of regulations with substantially the same content over a long period of time indicates that they were seldom actually enforced. Indeed, it might even be correct to say that these regulations were often meant only to provide an excuse to increase revenue by imposing fines on recalcitrant moneylenders.[4] Even in the possibly rare cases of actual enforcement, the relief was available to the peasantry only on a one-time basis. The problem, consequently, remained as intractable at the middle of the nineteenth century as it was during the early years. As some debtors of a village in Nuwakot district complained in 1854, "We pay Rs 200 on a loan of Rs 100, or thirty *pathis* of salt on a loan of one *muri* of rice, but the amount of the loan always remains the same as before."[5]

A high rate of interest does not in itself have an adverse impact on the economic condition of the peasant, if the loan is meant for a purpose that yields him a still higher benefit. In other words, the adverse impact of high rates of interest may be offset to a significant extent if the loan is used for productive purposes. There is evidence, however, that peasants in nineteenth-century Nepal usually incurred loans in the form of such basic necessities as "foodgrains, cloth, and money."[6] Even if amelioration of the condition of the debt-ridden peasantry was a serious objective of governmental policy, efforts to achieve that objective through regulation of rates of interest only tackled the symptoms and not the basic causes of the problem. There is no evidence that the government made any effort to improve the condition of the peasantry with the objective of reducing their dependence on moneylenders.

Impact of New Tax-Assessment and Collection Systems

This interpretation of official policy in respect to agrarian indebtedness is substantiated by the manner in which the government introduced innovations in the fields of tax assessment and collection that further tightened the squeeze on the peasant and made him even more vulnerable to indebtedness. The replacement of *adhiya* rents by *kut* rents in the central hill regions during the early nineteenth century, as discussed in Chapter 4, was one of these innovations. Notwithstanding the comparative advantages of fixed *kut* rents as secure variants, which tend to improve reserve allocation and productivity,[7] additional problems emerged for the small peasant. The cost variant was, no doubt, secure in absolute terms, but in terms of a percentage of the actual produce it was an extremely insecure variant. Chapter 4 had also discussed how the peasant obtained an automatic remission in rents under the *adhiya* system in the event of any damage to the land or crop failure. In contradistinction, even when production declined by one-fourth, no remission was given in *kut* rents. A partial failure of crops consequently left the peasant worse off under the *kut* system than under the *adhiya* system, and made him more vulnerable to indebtedness.

There was also a progressive trend toward the commutation of payments due on both rice-lands and homesteads, which meant the superposition of a money-tax system on a subsistence economy at a low level of monetization. For the small peasant, the chief way to raise cash was to sell that part of the produce that was payable in any case to the landlord. However, the village moneylender, who usually also dealt in agricultural produce, was the sole outlet for the conversion of the peasant's produce into cash, both because of the small quantities involved and the client-patron relationship between the two. The alternative way to obtain cash was to borrow from the moneylender. If the peasant was lucky and reaped a good harvest, he repaid the moneylender in kind. Otherwise, payments accumulated.

Nor was this all. Rice-lands yielded a crop only once a year, or, if wheat too was cultivated, at most twice a year. A *jagirdar* accordingly received an income from his lands only once or twice in a year. If the duties of his office required his presence at distant places where rents could not be transmitted to him during the months when crops ripened, he would even have to go without his rents. Such a situation occurred frequently during the early years of the nineteenth century, when many *jagirdars* belonging to the army were deputed to the front. In order to meet that difficulty, peasants who cultivated rice lands on *adhiya* or *kut* tenure were placed under the obligation of supplying loans to their *jagirdar*-landlords. Those loans were subsequently adjusted against the rents due to the *jagirdars* when crops were harvested. Peasants were allowed to charge interest on such loans at the rate of five per cent.[8] The system may have helped to remove the financial difficulties of *jagirdars*, but it increased the burden on the peasant. He had to raise large amounts of cash before his crops were ready for harvest, often at short notice. Moreover, he was allowed to charge an interest of only five per cent, although the village moneylender who supplied him with cash to meet this liability charged several times as much. There is evidence, nevertheless, that the obligation of peasants to supply loans to their *jagirdar*-landlords became defunct after the gradual introduction of the *tirja* system during the 1820s. *Jagirdars* were thereafter able to obtain payments in advance from the brokers to whom they sold their *tirjas*.

If tax assessment and collection policies made the peasant more vulnerable to indebtedness, measures aimed at increasing agricultural production through the reclamation of virgin lands made him no less so. The problem was prominent particularly in the Tarai region. There the government traditionally followed the policy of allotting large tracts of virgin lands to enterprising individuals for reclamation and colonization. Such individuals were expected to attract settlers from the adjoining Indian territories and provide loans and other facilities to them. Although there is very little direct information concerning the rates of interest charged on such loans, contemporary accounts of the adjoining Indian territories lead us to infer that the rates were usurious. In Purnea district of Bihar, which adjoins Morang in the eastern Tarai region of Nepal,

Francis Buchanan noted in 1809-10:

> On most estates it is customary to assist new tenants by a little money advanced. If he brings implements and cattle, the landlord or his agent advances grain for seed and food. The latter is paid back from the first crop, with an addition of 50 per cent; twice as much is required from the former. As the loan is seldom for more than six months, this is an enormous usury.[9]

Buchanan has also noted that "a large proportion of the farmers are in debt chiefly to merchants of various kinds who make advances for their produce."[10]

Slavery and Bondage

The foregoing sections show how the cumulated burden of indebtedness consequent to usurious rates of interest enhanced the economic burden borne by the peasant. The problem did not concern solely his financial condition, but affected even his personal liberty. When loans accumulated, the transaction often culminated with the purchase of the debtor by the moneylender. The debtor then became the moneylender's slave. Alternatively, the debtor worked for his creditor without wages in lieu of interest,[11] and thus became a bondsman. Bondage meant that "the creditor shall not demand interest, while the debtor shall not demand wages."[12] The system owed its origin also to the traditional taboo on Brahman landowners operating an ox-drawn plough with their own hands to till their fields.[13] They accordingly advanced interest-free loans to landless agricultural labourers, who were usually of low caste status, to do the work.

The prices at which poor people were sold as slaves may be regarded as reliable indicators of the economic condition of the lower strata of the peasantry in the hill regions of Nepal during the nineteenth century. These prices appear to have been determined by the low capitalized value of human labour as well as by demand and supply. In Jumla, a poor and mountainous region where the supply of slaves was possibly higher and the demand lower than in the capital town of Kathmandu, a slave could be bought for Rs 20[14] and an entire family, consisting of parents and two children, for Rs 80.[15] The official price of rice in Jumla at that time was seven *pathis* per rupee,[16] hence a slave girl was worth about

10½ *muris* of rice. In Kathmandu, on the other hand, "the price of slaves ranges for females from 150 to 200 rupees, and for males from 100 to 150 rupees."[17]

The size of the loans for which people were bonded, and the long periods of time for which they remained in bondage as a result of their failure to repay the loan, are additional economic indicators. In early 1865, Kathmandu received a complaint from a blacksmith in a village of Doti district that some local persons had wrongfully claimed his mother to be a slave and kept her in detention. The woman had been bonded by her father several years previously for a loan of Rs 6. The loan appeared to have remained unpaid for at least two decades.[18] At Barhabis in Jumla, a landless labourer remained a bondsman for six years for a loan of Rs 12. He was then freed by a local landowner and set up on a service-tenure allotment; the loan had not been repaid even 17 years later.[19]

For the moneylender, slaves and bondsmen were a source of cheap labour which he could use in various ways. In Kathmandu Valley, "most great proprietors . . . employ stewards with their servants and slaves, to cultivate some land for supplying their families."[20] In wealthy households, slaves were "generally employed in domestic work, wood-cutting, grass-cutting, and similar labour."[21] In the hill regions, wealthy *rakam* landlholders often deputed their bondsmen to transport mail and government supplies on their behalf. In one case, a moneylender who had been conscripted during the 1855-56 Nepal-Tibet war fulfilled his obligation by sending his bondsman instead.[22] Because of such economic and other benefits, the enslavement and bondage of the peasantry as a result of indebtedness appear to have been chronic problems in several parts of the country during the nineteenth century.[23]

At the same time, available evidence suggests that the institution of slavery, unlike bondage, was often not economic for their owners. Both slaves and bondsmen forfeited their labour power as a means of economic gain, but the creditor was under no obligation to provide a bondsman with means of subsistence. A slave, on the other hand, had to be fed, clothed, and sheltered. It was possibly because of the high costs of the slave's maintenance, compared with his low productivity, that owners in the western hill region often set up their slaves on small allotments.[24] This ensured them a cheap supply of labour without bearing the obligation of the slave's maintenance.

The foregoing sections discussed two ways in which a money-lender could get a return on his investment during the early years of the nineteenth century: collection of interest in cash or in kind, or the exploitation of the debtor's labour-power through slavery or bondage. Because the element of risk was high, neither arrangement was a satisfactory one from the viewpoint of the moneylender. Collection of interest was, at times, difficult because the debtor's crops might be destroyed by natural calamities, or the debtor himself might die or abscond. The liabilities of a deceased debtor or bondsman were assumed by his heirs, but a dead slave meant a complete loss of the owner's capital. High rates of interest, or harsh conditions of slavery or bondage, were, therefore, often no more than an attempt by the moneylender to provide insurance for his investment.

Land Mortgage

Moneylenders usually seek to avoid such risks by advancing loans against the security of the peasant's land. This can be done in either of two ways: through a simple or possessory mortgage. A simple mortgage means that the borrower assumes the obligation to pay interest on the loan, empowers the creditor to forfeit the plot of land mentioned in the bond, and meanwhile, cultivates the land himself. Under the system of possessory mortgage, on the other had, the borrower does not stipulate any interest, but surrenders his land to the creditor, who then cultivates it personally or appoints a tenant, often the borrower himself. Mortgage entitles a moneylender to collect an income from lands which he does not actually own or cultivate. Consequently, moneylenders become yet another category of interests, along with the landowning and village elites, who claim a share in the crops the peasant harvested.

The system of possessory mortgage, in particular, causes economic hardships to the peasant because it deprives him of a source of income with which he could pay back the loan. Even if he continues to cultivate the mortgaged land in the capacity of a tenant, he is compelled to pay at least half of the produce as rent to his moneylender-landlord. In other words, the peasant's financial obligations increases, whereas his income declines. Possessory mortgages, therefore, often long remain unredeemed. Of greater importance is the ratio between the size of the loan and the area of land mortgaged. If a moneylender acquires a large area of land on mortgage on a small loan, the income he can get from the land means a very high rate of return on his investment.[25] From the borrower's point of view, this means an unduly high rate of interest. At the same time, the system makes it unnecessary for the borrower to undertake the obligation of paying interest, and even enables him to redeem the mortgage whenever he can acquire the means to do so.

It is possible that the borrower fares worse under the system of simple mortgage, because he then remains under the obligation of making interest payments regularly. In the event of default in such payments, the moneylender can add up the arrears and renew the bond, in effect charging compound interest. When arrears of payment accumulate to a level which the moneylender judges to be approximate to the value of the mortgaged land, he may compel the borrower to surrender the land, thereby cancelling the loan. The moneylender then augments his land holdings, whereas the borrower becomes a landless labourer.

Obviously, the debtor can mortgage his land to his creditor only if his rights in that land are secure either by law or by custom that has gained administrative sanction. At the middle of the nineteenth century, not all categories of individual rights in land enjoyed that status; hence the scope for mortgage was usually limited to *birta*, *kipat* land in the eastern hill region, and *raikar* lands in the Baisi region. Available evidence shows that possessory mortgage transactions were quite common in these categories of lands during the early nineteenth century. Indeed, the steps that the government initiated from time to time to mitigate the adverse effects of possessory mortgages on the economic condition of the peasantry indicate the acuteness of the problem. In 1809, for instance, all lands mortgaged in the eastern hill region were restored to their owners, and the debts were cancelled, if the creditor had already collected the amount originally loaned.[26] In 1844, moneylenders in Jumla were similarly ordered not to enslave their debtors, or dispossess them of their lands.[27] We may assume that these orders were ignored by moneylenders. In fact, moneylenders at times behaved in such a high-handed fashion that they refused to restore the mortgaged lands even when the debtor repaid the loan, or to accept repayment and restore the mortgaged lands to the debtor.[28]

Inasmuch as occupancy rights in *raikar* land in the central hill region were not transferable during the early years of the nineteenth century, mortgages of the type common in the *kipat* areas of the eastern hill region and in the Baisi region were usually unknown. Nevertheless, there is evidence that extralegal mortgage transactions were quite common in that region on *jagir, rakam,* and other categories of *raikar* lands. In 1832, for instance, the government received reports that a large number of mail-carriers employed under the *rakam* system at Thankot in Kathmandu had mortgaged their lands to moneylenders and left the village.[29] There is evidence that a similar problem affected other parts of the central hill region as well.[30]

Rana Policies

Agrarian indebtedness, and the consequent impoverishment, enslavement, or bondage of the peasant, or the loss of his rights in the lands he tilled, appear to have become acute problems at the time of the commencement of Rana rule. In view of their adverse impact on agricultural production and stability of the agrarian population, the Rana government initiated a number of measures aimed at mitigating the burden of indebtedness, regulating enslavement and bondage, and ameliorating the condition of slaves and bondsmen.

The 1854 legal code reiterated the traditional ban on the collection of interest in excess of 10 per cent yearly. It prescribed that on accumulated loans only double the amount of the principal need be paid after 10 years. The maximum that could be paid on accumulated loans in kind after 10 years was three times the principal amount. Another provision was that all payments of interest made in excess of 10 per cent in the past should be deducted from the principal amount.[31] The code, in addition, contained provisions permitting a debtor to declare himself bankrupt with the consent of his creditors. In that event, his assets were utilized to settle their claims in proportion to the amount repayable to each. The creditor was given the choice of accepting repayment whenever the debtor became able to do so.[32]

Neither of these measures appears to have been overly effective in mitigating the burden of agrarian indebtedness. For the debtor, the pressure of the moneylender was always more tangible and effective than the force of the law. Nor were the legal provisions practical. For instance, provisions relating to rates of interest were subject to the condition that documentary evidence of the loan was available. The condition was obviously inapplicable in cases of agrarian indebtedness. Similarly, it is doubtful that the facility of declaring oneself bankrupt was actually available to agrarian debtors.

The Rana government also attempted to regulate indiscriminate enslavement and bondage. Several castes and communities were declared immune from enslavement. These included members of those Newar groups from whose hands high-caste people were permitted to take water,[33] and the Limbu communities of the far-western hill region.[34] Slaves could thereafter be procured only from liquor-drinking or untouchable castes and communities.[35] Members of these castes and communities too could be enslaved only if they were guilty of such crimes as incest,[36] burglary, and infanticide.[37]

In 1852, the Rana government banned the sale of free men in the Baisi region; only existing slaves and their children could thereafter be bought and sold.[38] Inasmuch as the measure were not retroactive, the condition of existing slaves remained unchanged. This lacuna was partially filled up in the Baisi region during 1853-54, when all persons enslaved after the commencement of Rana rule in 1846, and children born of slaves were declared free.[39] These provisions received countrywide application through the 1854 legal code, which imposed a general ban on the enslavement of defaulting debtors and other free men.[40] The code also abolished the right of parents to sell their children into slavery.[41]

Because of these reforms, indebtedness no longer remained a cause of slavery. In 1851, Captain Orfeur Cavanagh noted that "many slaves are born free, being the children of parties in necessitous circumstances and sold by their parents."[42] About a quarter-century later, Dr Daniel Wright, writing on the same problem, did not mention indebtedness as a cause of slavery. He noted that people could be enslaved only "for certain crimes, such as incest and some offences against caste."[43]

In addition, the Rana government initiated several measures to ameliorate the condition of slaves and bondsmen. Slaves traditionally enjoyed the right to own personal property.[44] The legal code prescribed that such property should not be confiscated even when slaves committed offenses ordinarily punishable through confiscation of property. Slaves could also inherit property, and even received priority in subdivision of property if their coparcencers were free men. Owners could, of course, emancipate their slaves, but the legal code also envisaged partial freedom, under which ex-slaves continued working for their masters but could not be sold to others.[45] The legal code, in addition, prescribed the value of slaves for the restitution of claims on the basis of sex and age. The highest price of Rs 120 was fixed for a slave girl in the 12-40 age group, while a male slave of the same category was worth Rs 20 less. Prices were lower for slaves who were below 12 years or above 40 years of age. A 60 years old slave woman fetched only Rs 50.[46] Although these prices were not applicable to actual transactions in slaves, it is possible that they set the minimum limits.

The 1854 legal code similarly contained a number of provisions aimed at making the bondage system less oppressive. Bondage was permitted only with official sanction.[47] The bondage of children under 16 years of age was prohibited. Persons of above this age were declared free if they complained that they had been bonded forcibly by their parents. Relatives of a dead bondsman were under no obligation to assume his debts, and even his son could be held as a bondsman only if this had been so stipulated previously in writing.[48]

Nevertheless, these were only palliative measures aimed at making the institutions of slavery and bondage less oppressive and inhuman; the objective was obviously not to abolish them. The 1854 legal code actually buttressed the authority of owners over their slaves. Disobedient slaves could be given physical punishment, or else put in irons or shackles and kept in confinement, albeit with official approval. However, no such approval was necessary in the case of runaway slaves.[49] Persons who helped slaves to escape were liable to pay compensation to the owners.[50] The property of runaway slaves accrued to their owners.[51]

Bondage too remained a lawful institution, and the government even provided assistance to moneylenders in capturing runaway bondsmen.[52] If a runaway bondsman was recaptured, he was forced to pay compensation for his creditor's loss of his services, subject to a limit of double the amount of the original loan. Alternatively, he could be put in fetters and kept in detention for a term which varied according to the amount of the loan.[53]

Rana efforts to regulate the systems of slavery and bondage, and to ameliorate the condition of slaves and bondsmen, nevertheless left untouched the basic role of the moneylender as one of the several claimants to the peasant's crop. In fact, several other developments occurred in the agrarian field after the commencement of Rana rule that reinforced that role of the moneylender. The manner in which a market in land emerged in *raikar* land in all parts of the country as a result of the agrarian reforms undertaken by the Rana government during the third quarter of the nineteenth century has already been discussed in Chapter 8. Thanks to that development, *raikar* lands in all parts of the country were mortgaged on a growing scale. Paradoxically, the gradual strengthening of the peasant's rights in land made him more vulnerable to expropriation by moneylenders.

In these circumstances, it is not surprising that restrictions on the enslavement and bondage of debtors had little practical effect on the moneylender's operations. Rather, he was now able to undertake mortgage transactions in land with greater confidence and security. More important, he was often even able to foreclose mortgages and assume legal ownership of the debtor's land.

Proletarization and Emigration

There seems little doubt that agrarian indebtedness was the main factor responsible for the progressive proletarization of the small peasant in Nepal during the nineteenth century. Even if he was not actually expropriated from his holding, the burden of debt increased from year to year at compounded rates of interest and was often so heavy as to remain a permanent legacy. Consequently, even peasants who held the lands they tilled in their own names were little better than tenants or bondsmen.

The nineteenth century witnessed a large-scale emigration of people from the hill areas of Nepal to Bengal, Assam, Burma and elsewhere. There is no doubt that a number of factors were responsible for this phenomenon, such as growing population and the consequent pressure on the available agricultural land. There is evidence, at the same time, that a sizeable number of these emigrants consisted of slaves and debtors who were harassed by their owners and creditors, and expropriated from their lands.[54] The economic loss that Nepal incurred as a result of such large-scale emigration of its economically active labour force is indicated by the contributions the emigrants made in the development of the Indian economy. The development of the coal-mining industry in

the adjoining provinces of Bihar and Bengal, and of the tea industry in Bengal and Assam, during the third quarter of the nineteenth century created a big demand for unskilled labour which, to
some extent, was met through Nepali emigrants.[55] These emigrants
also made a significant contribution in opening up new areas for
reclamation and settlement in India. They have been described by
a contemporary European observer as:

> ...industrious and enterprising cultivators, greatly superior to
> the other races in this quarter, and destined to do more and
> more for the settlement and colonization of these hills. They are
> the men who break up the land with the plough, and show the
> other races how to give up the barbarous method of tillage with-
> out it.[96]

There can perhaps be no better words to illustrate the basic inequities of Nepal's agrarian system, which compelled the nation to
export an important economic asset in this manner.

The main consequence of agrarian indebtedness in nineteenth-
century Nepal in the context of the present study was the emergence
of a class of moneylender-cum-landlords in the village who were
interested in agriculture only to the extent that they might draw an
income from it. Along with the landowning and village elites, these
groups too claimed a share in the peasant's surplus produce. To
be sure, the moneylender performed the essential task of credit
supply in an economy where alternative sources were virtually non-
existent. Even then, the interest that he charged was high enough
to provide insurance for his risks. The peasant was, consequently,
forced to pay the premium for insuring the moneylender against
risks due primarily to the heavy burden of taxes and other payments exacted from him by the landowning and village elites.

NOTES

[1]Ministry of Law and Justice, "Swasti Shri 9 Maharajadhiraja Ramasahabata
Bandhi Baksyako Thiti" (Regulations promulgated by His Majesty King Rama
Shah), secs. 4-5, in *Shri 5 Surendra Bikram Shahdevaka Shasanakalama Baneko*

Muluki Ain (Legal code enacted during the reign of King Surendra Bikram Shah Deva), Kathmandu, Ministry of Law and Justice, 2022 (1965), pp. 695-96. For a full translation, see *Regmi Research Series*, Year 2, No. 2, February 1, 1970, pp. 49-50.

[2]Mahesh C. Regmi, *A Study in Nepali Economic History, 1768-1846*, New Delhi, Manjusri Publishing House, 1971, pp. 98-99, 191-92.

[3]Dinesh Raj Pant, "Swami Maharaj Rana Bahadur Shahko Vi Sam. 1862 Ko Bandobast" (Administrative arrangements made by Swami Maharaj Rana Bahadur Shah in A.D. 1805), sec. 33, *Purnima*, 24, Magh-Chaitra 2027 (January-March 1971), pp. 238-67 For a full translation, see *Regmi Research Series*, Year 3, No. 6, June 1, 1971, pp 128-37.

[4]"Order Regarding Punishment of Usurious Moneylenders in the Eastern Hill Region," Chaitra Badi 4, 1863 (March 1807), *RRC*, Vol. 6, p. 809.

[5]"Order regarding Complaint of Debtors in Panchsayakhola," Poush Sudi 3, 1911 (December 1854), *ibid.*, Vol. 33, p. 214.

[6]"Complaint regarding Forced Enslavement of Bondsman in Doti," Falgun Badi 2, 1971 (February 1865), *ibid.*, Vol. 21, pp. 554-55.

[7]Clive Bell, "Ideology and Economic Interests in Indian Land Reform," in David Lehmann, *Agrarian Reform & Agrarian Reformism*, London, Faber and Faber, 1974, pp. 198-99.

[8]Regmi, *Nepali Economic History, op. cit.*, p. 98.

[9]Francis Buchanan, *An Account of the District of Purnea in 1809-10*, Patna, Bihar and Orissa Research Society, 1928, p. 436.

[10]*Ibid.*, p. 435.

[11]Cf. "Order to the Dwares, etc. of Pinda, Kiranchok," Chaitra Sudi 10, 1906 (March 1850), *RRC*, Vol. 64, pp. 654-56.

[12]"Complaint of Mangle Sarki of Liglig, Gorkha," Magh Sudi 8, 1921 (January 1856), *ibid.*, Vol. 21, pp. 328-31.

[13]"Order to Dwares, etc. of Dhulikhel to Refund Fines Imposed on Upadhyaya Brahmans for Drawing the Plough," Chaitra Badi 10, 1912 (March 1856), *ibid.*, Vol. 56, pp. 396-97.

[14]"Receipt for Rs 20 as the Price of Jyuna Chudara, a Slave," Baisakh Sudi 2, 1903 (April 1846), in Yogi Naraharinath (ed), *Itihas Prakash* (Light on History), Vol. 2, Bk. 2, Kathmandu, Itihas Prakashak Sangha, 2013 (1956), p. 253.

[15]"Sale-Deed for Slaves by Chautara Birajit Shahi of Jajarkot," Bhadra Sudi 2, 1929 (August 1872), in Naraharinath, *op. cit.*, Vol. 2, Bk . 1, p. 341.

[16]"Miscellaneous Regulations for Jumla," Falgun Badi 30, 1915 (February 1859), *RRC*, Vol. 29, pp. 272-79.

[17]Daniel Wright (ed.), *History of Nepal* (Reprint of 1877 ed.), Kathmandu, Nepal Antiquated Book Publishers, 1972, p. 45.

[18]"Complaint regarding Forced Enslavement of Bondsman in Doti," 1865, (See n. 5 above).

[19]"Complaint Regarding Refusal of Freed Bondsman to Repay Loan in Jumla," Falgun Badi 7, 1921, (February 1966), *RRC*, Vol. 21, pp. 546-47.

[20]Francis (Buchanan) Hamilton, *An Account of the Kingdom of Nepal*, (Reprint of 1819 ed.), New Delhi, Manjusri Publishing House, 1971, p. 220.

[21]Wright, *op. cit.*, p. 45; Orfeur Cavanagh, *Rough Notes on the State of*

Nepal, Calcutta, W. Palmer, 1851, p. 95.

[22]"Complaint of Kalu Lama, a Bondsman," Falgun Badi 7, 1918 (February 1862), *RRC*, Vol. 47, p. 743.

[23]Regmi, *Nepali Economic History, op. cit.*, pp. 117-23.

[24]"Order regarding Land Allotments made to Slaves in Doti," Kartik Badi 4, 1912 (October 1855), *RRC*, Vol. 56, pp. 530-33.

[25]Cf., according to the 1888 Legal Code: "In case a peasant cultivating *raikar* lands ... obtains a loan (from a creditor), the latter may complain: 'My holding, which can produce much (grain), is being cultivated (by the creditor) on payment of a small sum of money. I do not have the capacity to pay back the loan'," Government of Nepal, "Jagga Pajaniko" (On land evictions), sec, 38, in *Muluki Ain* (Legal Code), Kathmandu, Biradev Prakash Press, 1945 (1888), pt. 3, p. 34.

[26]"Judicial Regulations for Areas East of the Dudhkosi River," Marga Badi 9, 1866 (November 1809), *RRC*, Sec. 12, Vol. 40, p. 127.

[27]"Thekthiti Arrangements in Gam, Jumla District," Chaitra Sudi 7, 1900 (April 1844), in Narahirinath, *op. cit.*, Vol. 2, Bk. 2, p. 278.

[28]"Judicial Regulations for Areas East of the Dudhikosi River," Marga Sudi 5, 1864 (December 1807), *RRC*, Sec. 12, Vol. 6, p. 956; "Judicial Regulations for Areas between Sunkosi and Dudhkosi Rivers," Poush Badi 8, 1894 (December 1837), *ibid.*, Sec. 10, Vol. 26, p. 699; "Complaint against Refusal of Creditor to Redeem Mortgaged Lands in Humla," Falgun Badi 2, 1921 (February 1865), *ibid.*, Vol. 21. pp. 557-60.

[29]"Order to the Dwares of Thankot Village," Magh Sudi 4, 1888 (January 1832), *ibid*, Vol. 45, p. 11.

[30]"Order regarding Land Allotments to Hulaki Porters in Bhirkot," Poush Badi 3, 1919 (December 1862), *ibid.*, Vol. 62, p 396-97

[31]Ministry of Law and Justice, "Sahu Riniko" (On creditors and debtors), Sec. 1, in *Shri 35 Surendra . . . Muluki Ain*, p. 92.

[32]Ministry of Law and Justice, "Damasahiko" (On insolvency), *ibid.*, pp. 109-11.

[33]Ministry of Law and Justice, "Misa Khatko" (Sexual offenses among the Newar Community), Sec., 1, *ibid.*, p. 644.

[34]"Ban on Enslavement of Limbus," Magh Sudi 9, 1917 (January 1861), *RRC*, Vol. 33, pp. 276-77; "Ban on Enslavement of Khambus," Magh Sudi 9, 1917 (January 1861), *ibid.*, Vol. 49, pp. 173-74.

[35]Ministry of Law and Justice, "Masinya Jiu Amalile Linako" (Amali's rights over enslaved persons), Sec. 1, in *Shri 5 Surendra . . . Muluki Ain*, p. 367.

[36]Ministry of Law and Justice, "Masinya Matuwaliko" (On the enslavement of members of liquor-drinking communities), Sec. 3, *ibid.*, p. 548.

[37]"Administrative Regulations for the Bheri-Mahakali Region," Kartik Sudi 8, 1908 (October 1851), *RRC*, Sec. 18, Vol. 49, p. 95.

[38]*Ibid.*, Sec. 13, Vol. 49, p. 91; "Order Regarding Thek-Thiti Arrangements in Chinga Dara, Dullu-Dailekh," Marga Badi 5, 1938 (November 1881), *ibid.*, Vol. 58, p. 137.

[39]"Order Regarding Emancipation of Slaves Enslaved in the Region West of the Bheri River After A.D. 1847," Baisakh Sudi 14, 1911 (April 1858), *ibid.*,

Vol. 49, pp. 272-72.

[40]Ministry of Law and Justice, "Jiu Masnya Bechanaya," (On traffic in human beings), Sec. 1, in *Shri 5 Surendra . . . Muluki Ain*, p. 355.

[41]Ministry of Law and Justice, "Kamara Kamari Bechatato" (On transactions in slaves), Sec. 3, *ibid.*, pp. 352-53.

[42]Cavanagh, *op. cit.*, p. 95.

[43]Wright, *op. cit.*, p. 45.

[44]"Order Regarding Property of Runaway Slave in Isma, 1901 (1844), *RRC*, Vol. 3, p. 33.

[45]"On Traffic in Human Beings." (See n. 40 above).

[46]*Ibid.*, Sec. 4, p. 356.

[47]*Ibid.*, Sec. 12, p. 358; "On Transactions in Slaves." Sec. 4. (See n. 41 above).

[48]"On Traffic in Human Beings," Secs. 6-7, p. 357. (See n. 40 above).

[49]Ministry of Law and Justice, "Bandha Kamara Bhagaunyako" (On helping bondsmen and slaves to escape), Sec. 7, in *Shri 5 Surendra . . . Muluki Ain*, p. 350.

[50]*Ibid*, sec. 1, p. 349; "Order to Local Functionaries in Khinchet Regarding Assistance to Bondsmen and Slaves to Escape," Marga Sudi 8, 1909 (December 1852), *RSC*, Vol. 29, pp. 241-42.

[51]"Order Regarding Jahar Singh Mijhar's Claim to His Slave's Property," Falgun Sudi 15, 1912 (March 1856), *ibid.*, Vol. 56, pp. 667-70.

[52]"Sadar Dafdarkhana's Order to Thekdar Chandrabir Adhikari Chhetri of Panchasaya Khola," Bhadra Sudi 9, 1920 (September 1863), *ibid.*, Vol. 55, p. 8; "Order to Amalis in Kandrang, Chisapani, Makwanpur, etc. Regarding Capture of Runaway Bondsmen and Slaves," Magh Badi 2, 1906 (January 1847), *ibid.*. Vol. 64, pp. 541-45.

[53]"On Helping Bondsmen and Slaves to Escape," Sec. 9, p. 351. (See n. 48 above).

[54]In 1830, for instance, local functionaries in the Solukhumbu area were ordered to persuade debtors "who have crossed the mountains for fear of their creditors" to come back. "Royal order to the Ganbas and Mijhars of Pargaghat," Bhadra Badi 14, 1887 (August 1830), *RRC*, Vol. 34, p. 77; "Order to Amalis in Kandrang . . . ," 1847, (see n. 52 above); "Order Regarding Emigration of Limbus," Ashadh Badi 11, 1907 (June 1850), *ibid.*, Vol. 64, p. 723.

[55]Cf. Maharaja Chandra Shum Shere Jung Bahadur Rana, *Appeal to the People of Nepal for the Emancipation of Slaves and Abolition of Slavery in the Country*, Kathmandu, Suba Rama Mani A.D., 1925, p. 31.

[56]Richard Temple, *Journals Kept in Hyderabad, Kashmir, Sikkim and Nepal*, London, W.H. Allen, 1887, Vol. 2, p. 196.

9

Agricultural Development Policies

The previous chapters dealt with the different categories of elite groups who had a claim in the peasant's produce, and the institutional mechanism through which the claim was actually enforced. The objective of the system was to maximize the amount of economic surplus extracted from the peasantry within the limits of the current techniques and volume of production. In other words, the agrarian institutions that the Rana rulers and their predecessors devised during the nineteenth century were designed primarily to enable the government to augment its revenue, as well as the income of the landowning elite, rather than to increase agricultural production. Such an increase, which helped the government to widen the tax-base, was the objective of another set of measures, which will form the subject matter of this chapter.

Efforts to Extend the Cultivated Area

Agricultural production can be increased either by increasing productivity per unit of the cultivated area, or by extending the area under cultivation. Nineteenth-century Nepal faced no real choice as between these two alternatives. Productivity can be increased through the application of more labour and capital, both of whom were in short supply, whereas land was relatively plentiful. Measures to increase agricultural production were, consequently, confined for the most part to an extension of the area under cultivation.

Efforts to extend the cultivated area during the nineteenth century were concentrated mainly in the Tarai region. The reasons are obvious. This region possessed considerable potential for development because of relatively low density of population and extensive tracts of cultivable lands and forests. Proximity to the markets of northern India increased the commercial value of its timber and other natural resources and also facilitated trade in agricultural commodities. Such favourable factors had led to efforts to develop the Tarai region as early as during the closing years of the eighteenth century. The observation made by Hamilton in 1814, that "the Gorkhalese have cleared much of the country" in the Tarai region,[1] bears testimony to the success of these efforts.

Conditions during the latter part of the nineteenth century were unusually favourable for accelerating the pace of agricultural development in the Tarai region. As noted in Chapter 2, more stable and friendly relations were established during that period with the British Indian government. Traditional constraints in the colonization of the Tarai region, which had already weakened after the advent of Gorkhali rule, became obsolete. Nor was this all. The latter part of the nineteenth century also witnessed a big spurt in economic activity in northern India, mainly because of the development of railway transport facilities. This inevitably had spread effects on the Tarai region of Nepal. By the end of the nineteenth century, India's railway network had touched the Nepal-India border at several points. The construction of railway tracks led to an increased demand for construction materials, such as timber and boulders, which were readily and abundantly available in the Tarai region of Nepal.[2] Of greater importance was the fillip the new transport facilities gave to the production and export of such agricultural commodities as rice and jute from the Tarai region. These developments opened up unprecedented prospects for agricultural expansion in the Tarai region. The Rana rulers took prompt advantage of the situation, inasmuch as the land and other natural resources of that region constituted a major source of the income not only of the landowning elite but also of the government.

Land-Reclamation Policies

Measures to extend the area under cultivation formed the main plank of the agricultural development policy of the Ranas. The traditional policy of encouraging private enterprise in this field through fiscal concessions[3] formed part and parcel of these measures. The legal code, therefore, prescribed that any person who brought under the plough lands adjoining cultivated holdings should be granted tax-exemption for three years in the hill region, and for five years in the Tarai region.[4]

Tax concessions alone could not provide a sufficient incentive for the reclamation of waste lands. The benefits usually went to individual farmers who would have reclaimed lands of convenient location even in the absence of such concessions in order to meet the needs of a growing family. Available evidence also suggests that the fiscal concessions granted by the government to encourage land reclamation through small-scale individual effort were frequently abused. The result was that though the government relinquished revenue through tax concessions, there was little net addition to the cultivated area.[5] In any case, small-scale individual enterprise could not provide the capital and entrepreneurial ability needed to open up the vast tracts of reclaimable forest and other lands then available in the Tarai region.

The main thrust of land reclamation policy in the Tarai region throughout the nineteenth century was, therefore, to encourage private enterprise in the colonization of large tracts of forests and other uncultivated lands whose development lay beyond the capacity of the local farmers because of inconvenient location or paucity of capital.[6] The Rana government assigned the role of such colonization to *jimidars*. Chapter 5 had described the *jimidari* system from the viewpoint of tax-collection, but the role of the *jimidar* as an agent of agricultural expansion in the Tarai region during the nineteenth century is no less important.

Legislation was, therefore, enacted to provide additional concessions and privileges to individuals who undertook colonization schemes on a large scale. According to the legal code:

If any person obtains an allotment of virgin lands anywhere in the Kingdom of Gorkha after 1852, and reclaims and irrigates such lands through his own labour and resources, he shall be granted as *birta* lands yielding an income of one rupee for each eleven rupees of additional revenue collected. He shall also be granted tax exemption for three years on the remaining area, and permitted to hold it on a taxable and inheritable basis from the fourth year. He may be granted tax-exemption even for five years or ten years, provided he does not seek one-tenth of the reclaimed lands as *birta*.[7]

More liberal regulations were promulgated for the eastern Tarai region during the 1860s with the same objective:

> Any person who undertakes to settle new *moujas* on uncultivated tracts of land between the Narayani and Mechi rivers in the Tarai region shall be allotted such lands. No taxes shall be collected from him for the first ten years. At the end of that period, one *bigha* of land shall be granted as *birta* for each eleven *bighas* reclaimed by him, and taxes shall be assessed on the remaining area. His rights in the reclaimed lands shall be inheritable, and the allotment shall not be cancelled even if he commits any crime.[8]

The colonizer was expected to procure settlers from India, or from tax-free lands in the district, and provide them with permanent allotments with full tax-exemption for the first five years. He was also required to supply necessary credit at statutory rates of interest. Such liberal concessions and privileges were necessary to compensate the initial overhead investments made in clearing forests, digging irrigation channels, building huts, and procuring bullocks and agricultural implements.

The Role of the State

For the government, *jimidari* was a useful and effective institution in implementing the policy of agricultural development in the Tarai region. The system provided full scope for private enterprise and capital investment without imposing any liability on the government. Because of these advantages of the *jimidari* system, colonization schemes were often executed under the direct auspices of the government only in forest lands comprising valuable timber whose export would fetch large amounts of revenue. The forests were then cleared, and the lands allotted to individual settlers, under the direct supervision of the government. Such schemes were executed in Morang in 1883[9] and in Nepalgunj in 1897.[10] Revenue from the sale of timber was the chief objective of these projects, but their impact on the agricultural development of the Tarai cannot be overlooked. No information is available about the execution of the projects, but there seems little doubt that they were largely successful, in view of the prospects of immediate profit from the sale of timber, and the favourable location of the project areas close to the Indian borders.

The development of irrigation facilities was another field in which the government often took a direct initiative, inasmuch as such facilities were of key importance in ensuring the success of land reclamation and settlement policies. The general policy in the eastern Tarai region was that the government would meet half of the cost of irrigation facilities constructed by local farmers or *jimidars*.[11] The government appears to have taken direct initiative in the construction of irrigation schemes in that region only during the late 1860s, possibly as an aftermath of the widespread drought and famine which had ravaged that region a few years previously. Such schemes were then taken up in several districts of this region under the supervision of the army, and local functionaries were placed under the obligation of raising half of the costs of construction from farmers whose lands would get irrigation facilities from the canals.[12] In the far-western Tarai districts of Banke, Bardiya, Kailali and Kanchanpur, local authorities were instructed to make arrangements for the repair of existing irrigation channels if such repair was beyond the capacity of the local people. They were also authorized to construct new irrigation projects at government expense if additional lands could be reclaimed through such facilities.[13]

Although these measures look impressive on paper, it is difficult to say that they rendered any permanent contribution in protecting crops in the Tarai region from the vagaries of the monsoon. There is little evidence to show that *jimidars* and other landlords in the Tarai ever took any real interest in agricultural production to spend much initiative and effort in undertaking the construction of irrigation schemes. It is similarly doubtful that local authorities ever took their duties seriously in such matters. Even in the rare cases of actual implementation, it is doubtful that the additional irrigation facilities were durable. Most of them were possibly nothing more than temporary channels and earthen embankments which did not outlast the first monsoon.

Immigration Policies

Measures aimed at encouraging land reclamation and settlement in the Tarai region could be implemented successfully only if adequate agricultural manpower was available and peasants were willing to go through the pioneering venture of breaking new soil and fighting against the lush vegetation of the tropical Tarai, malaria and other scourges, and the ravages of wild animals. However, the low density of population constituted the main constraint in the agricultural development of the Tarai region during the nineteenth century. Efforts to attract settlers to take up landholdings in the Tarai were, therefore, an important component of the land policy of the government. These efforts gathered further momentum after the emergence of Rana rule.

The problem of encouraging immigration into the districts of the Tarai region was not a simple one involving the transfer of population from the hill areas, because there also governmental policy aimed at stabilizing the rural population and checking the depopulation of homesteads. Moreover, pressure of population on the cultivable area available in the hills was not critical enough to induce migration to the Tarai, and the climate of the Tarai was unsuitable for hillsmen.

Since the early years of the nineteenth century, the government had tried to encourage hillsmen to occupy lands in the Tarai region, mainly through concessional rates of taxation.[14] The Rana government, in particular, also liberalized regulations relating to crime, slavery, and indebtedness as an additional inducement for hillsmen to shift to the Tarai. For instance, criminals and runaway slaves, who reclaimed waste lands, were entitled to pardon and freedom, while debtors were permitted to repay their loans in instalments.[15] These efforts appear to have met with some measure of success. In 1897, for instance, a part of the approximately 76,000 *bighas* of taxable lands in Kailali and Kanchanpur was being cultivated by 346 families from the hill region.[16]

Because of the difficulties in attracting settlers from the hill districts on a large scale, the Rana government continued the traditional policy of encouraging immigration from the adjoining areas of northern India also. Any Indian who moved into Nepali territory along with his family was given a free allotment of agricultural land, in addition to a homesite, and free supplies of building materials.[17] Such immigrants were even eligible for appointment as *jimidars*, although preference was naturally given to settlers from the hill areas of Nepal.[18] Non-resident Indians too were permitted to cultivate lands in the Tarai region of Nepal, subject to the condition that a local person stood surety for their tax liabilities.[19] These precautions were obviously necessary to prevent settlers from evading taxes "by escaping into British territory immediately after reaping the harvest."[20]

These "pull" factors from the Nepali side would possibly have remained ineffective in encouraging immigration into the Tarai region had not strong "push" factors operated from the Indian side during the nineteenth century. These "push" factors consisted of the insecurity of tenure from which the actual cultivator suffered in Bihar and Bengal, which adjoin the central and eastern Tarai districts of Nepal. The East India Company government had introduced new systems of land tenure and revenue collection in those provinces in 1793.[21] These systems, which became known as the permanent settlement, vested landownership rights in non-cultivating *zamindars*, whereas the actual cultivators were recognized as no more than tenants.[22] It was, therefore, impossible for them to acquire ownership rights in the lands that they tilled. Nor was this all. The actual cultivators were also denied the protection of the law in matters concerning the amount of rents they paid to the *zamindars* and the security of their tenure. Under the permanent settlement, therefore, the rights of cultivators deteriorated to the point where they become little more than tenants-at-will. An attempt was made to retrieve the situation in 1859, when legislation was enacted giving the right of occupancy to cultivators who had been in continuous occupation of the land for twelve years. For most cultivators, however, it was a difficult task to prove such continuous occupation.[23]

The situation was even worse in the Indian areas bordering the far-western Tarai districts of Banke, Bardiya, Kailali and Kanchanpur. After the 1857 rebellion, the British government recognized only the rights of non-working landlords, called *Talukdars*, and not those of the actual cultivators. The reason was that landlords wielded both power and influence to support the British rule, whereas ordinary peasants had neither.[24] Subsequently, steps were taken to protect the interests of the actual cultivators, so that they might not be "abandoned to the mercy of the Talukdars."[25] Persons who had been in continuous occupation of their holdings since 1844, that is, twelve years before the annexation of these territories by the British in 1856, were granted permanent rights, but all future accrual of occupancy rights was put to an end.[26] The situation became worse when the 1868 Oudh Rent Act conferred the right of occupancy on every tenant, "who, within thirty years next before February 13, 1856, was in possession as proprietor of some portion of land in a village." Such a tenant was given a heritable but not a transferable right of occupancy from August 24, 1866. No occupancy rights were recognized in the case of ordinary tenants. The result of these measures was that the majority of the peasantry were tenants and had no protection whatever against eviction or enhancement of rents.[27]

In view of the disabilities to which cultivators in the adjoining areas of India were thus subject, the tenurial and other facilities and concessions offered by the government of Nepal must undoubtedly have proved attractive. Chapter 7 discussed how tenurial rights were made secure in the Tarai region, and how holdings were eventually recognized as transferable. The right to transfer lands in this manner meant that settlers who had invested their labour and capital in financing reclamation schemes could sell these lands if they wanted to do so. This naturally reduced the risk and uncertainty involved in such schemes. Thanks to these policies, Indian immigrants were assured of land-allotments on liberal terms, with full prospects of legally-recognized ownership rights. Consequently, there was strong incentive for them to cross over to Nepal and obtain an allotment of waste land in which they had immediate assurances of a legal title. Available evidence indicates that settlers did come from India in large numbers.[28]

The Scale of Achievement

Although statistics of the area under cultivation in the Tarai region at different periods during the nineteenth century are not available, there seems little doubt that it recorded a significant expansion. During the 1850s, when the far-western Tarai had not yet become a part of Nepal, Oldfield had noted:

On the outskirts of the forest portions of jungle are, from time to time, 'cleared,' and the hitherto uncultivated land becomes absorbed into the open tarai. . . . By this process of 'reclaiming' land, which is constantly going on to a greater or less degree, not only does the forest become gradually diminished in its width, but the amount of land under cultivation in the open tarai is steadily but slowly increasing in extent, and consequently in value also.[29]

This observation is substantiated by the following statistics, which show that during the period from 1852 to 1862, land-revenue collections in the eastern Tarai region more than doubled:[30]

LAND-REVENUE COLLECTIONS IN THE EASTERN TARAI
REGION, 1852-62

District	1852 Rs	1862 Rs	Percentage of Increase
Morang	151,081	276,094	82.74
Saptari	181,582	357,921	97.11
Mahottari	174,025	352,467	102.53
Sarlahi	96,233	171,377	78.08
Rautahat	69,990	192,587	175.16
Bara	47,555	143,578	201.91
Parsa	43,676	92,777	112.42
Total Rs	764,142	Rs 1,586,801	107.65

The Rana government's land-reclamation policies were possibly even more successful in the far-western Tarai region, if only because the region was less developed than the eastern Tarai. For instance, an official report stated in 1867 that in Banke district, "new lands are being brought under the plough each year. Large tracts of forest lands have been cleared and converted into rice-lands through the construction of dams and canals."[31]

The Central Hill Region

Efforts to encourage private enterprise in extending the cultivated area were made in the hill region also. Reference has been made earlier to the provision made in the 1854 legal code to provide tax concessions and other facilities to persons who undertook large-scale land-reclamation schemes. A few such schemes appear to have been executed in the hill region also, particularly during the later years of the nineteenth century.[32] However, the hilly terrain provided relatively little scope for extending the cultivated area. Such extension, particularly in the case of rice-lands, usually meant terracing hillsides. The majority of peasants were seldom able to provide the costs of digging waste lands, and providing the additional seeds, equipment, and food required to finance large-scale reclamation schemes. Land reclamation in the central hill region was, therefore, largely limited to the growing needs of individual peasant families.

Multi-cropping is one way in which agricultural production can be increased in a situation where the scope for extending the cultivated area is relatively limited. Efforts were, therefore, made during the 1860s, albeit in a half-hearted fashion, to encourage the cultivation of wheat and barley in winter after the rice crop was harvested. During 1866-67, for instance, recurrent failure of crops set off a proposal to encourage the cultivation of such crops in Gulmi, Argha and Khanchi districts. The only measure taken to attain that objective was, however, the imposition of a ban on the grazing of cattle in the fields after the rice crop was harvested.[33] One wonders to what extent the ban was enforced, and whether big farmers, who owned most of the cattle in the village, actually refrained from using the rice fields of their poorer neighbours as pastures. In any case, the cultivation of winter crops does not appear to have been a successful experiment in the central hill region. Consequently, the ban on the grazing of cattle on rice-fields during winter was reimposed several times during the early years of the twentieth century.[34]

On the whole, it would appear correct to conclude that any increase in agricultural production that was achieved during the nineteenth century was due mainly to an extension of the cultivated area, particularly in the Tarai region. There is no evidence that there was any net increase in agricultural productivity or in the peasant's income. Given the relatively low density of population during the nineteenth century, policies aimed at extending the cultivated area were quite natural, but in the absence of any net increase in productivity or in the peasant's income, such extension alone did not contribute to economic growth.[35]

NOTES

[1]Francis Hamilton, *An Account of the Kingdom of Nepal*, (Reprint of 1819 ed.), New Delhi, Manjusri Publishing House, 1971, p. 64.

[2]Cf. "Order Regarding Export of Boulders from Banke District for Construction of Jumnaha Railway Line," Falgun Sudi 12, 1942 (March 1886), *RRC*, Vol. 54, pp. 678-83.

[3]Cf. Mahesh C. Regmi, *A Study in Nepali Economic History, 1768-1846*, New Delhi, Manjusri Publishing House, 1971, p. 144.

[4]Government of Nepal, "Jagga Birhaunyako" (On land reclamation) in *Ain* (Legal Code), Kathmandu, Manoranjan Press, 1927-28 (1870-71), Sec 22, p. 20; "Regulations for the Sadar Dafdarkhana Office," Magh Badi 1, 1919 (January 1863), *RRC*, Sec. 12, Vol. 47, p. 415.

[5]In Morang district, for instance, peasants often vacated their holdings and left for India during the monsoon, but subsequently returned to Nepal and accepted new allotments on concessional terms. *Regmi Research Series*, Year 7, No. 11, November 1, 1975, p. 218. Regulations were, therefore, promulgated for the Tarai region which denied fiscal concessions to peasants who vacated their old holdings and obtained new allotments. "Order regarding Land Reclamation in Kailali District" Kartik Badi 12, 1954 (October 1897). *RRC*, Vol. 61, p. 352; "Order regarding Fiscal Concessions on Newly-Reclaimed Lands in the Eastern Tarai districts," Shrawan Badi 11, 1919 (July 1862), *ibid.*, Vol. 29, pp. 452-64.

[6]"Revenue Regulations for the Naya Muluk Region," Marga Badi 6, 1918 (November 1861), *ibid.*, Sec. 71, Vol. 47, pp. 470-71; "Revenue Regulations for the Eastern Tarai Region," Marga Badi 6, 1918 (November 1816), *ibid.*, Sec. 68, Vol. 10, pp. 50-51.

[7]Government of Nepal, "Jagga Birhaunyako" (On land reclamation), in *Ain* (Legal Code) (1870-71 ed.), *op. cit.*, sec., 18, p. 153.

[8]"Land-Reclamation Regulations for the Eastern Tarai Region," Magh Badi 3, 1921 (January 1865), *RRC*, Vol. 21, pp. 115-17.

[9]"Order regarding Forest-Land-Reclamation Project in Morang," Poush Badi 9, 1942 (December 1885), *ibid.*, Vol. 54, pp. 177-81.

[10]"Regulations regarding Forest-Land-Reclamation Project in Nepalgunj," Kartik Sudi 12, 1954 (November 1897), *ibid.*, Vol. 61, pp. 432-56.

[11]Revenue Regulations for the Eastern Tarai region, 1861, Sec. 69, pp. 52-53. (See n. 6 above).

[12]"Order regarding Construction of Canals in the Eastern Tarai Districts," Poush Badi 6, 1923 (December 1866), *ibid.*, Vol. 63, pp. 488-91.

[13]"Survey Regulations for the Naya Muluk Region," sec. 25. Kartik Sudi 15, 1917 (November 1860), *ibid.*, Vol. 47, p. 430.

[14]"Order Regarding Concessional Land-Tax Rates for Hillsmen in Bara and Parsa Districts," Jestha Sudi 6, 1867 (May 1810), *ibid.*, Vol. 39, p. 213. These concessions were first introduced in 1793.

[15]Ministry of Law and Justice, "Jagga Jaminko" (On land), in Ministry of law and Justice, *Shri 5 Surendra Bikram Shahdevaka Shasan Kalma Baneko Muluki Ain* (Legal code enacted during the reign of King Surendra Bikram

Shah Dev), Kathmandu, the Ministry, 2202 (1965), Sec. 64-65, p. 36. After the British Indian government restored the Naya Muluk region to the government of Nepal in 1860, a law was enacted to grant amnesty to criminals slaves, etc. in the Naya Muluk Region. "Amnesty to Criminals and Runaway Slaves in the Naya Muluk Region," Aswin Sudi 3, 1917 (September 1860), *RRC*, Vol. 33, pp. 382-84.

[16]"Order Regarding Land Reclamation in Kailali District," Kartik Badi 12, 1954 (October 1897), *ibid.*, Vol. 61, p. 352.

[17]"Revenue Regulations for the Eastern Tarai Districts," Sec. 68, pp. 50-52. (See n 6 above).

[18]*Ibid.*, sec. 52, p. 30.

[19]*Ibid.*, sec. 24, pp. 17-19; "Revenue Regulations for the Naya Muluk Region," sec. 26, pp. 451-52. (See n. 6 above).

[20]Pudma Jang Bahadur Rana, *Life of Maharaja Sir Jung Bahadur Rana of Nepal* (reprint of 1909 ed.), Kathmandu, Ratna Pustak Bhandar, 1974, p. 254.

[21]B. H. Baden-Powell, *Land Systems of British India*, (Reprint) Delhi, Oriental Publishers, 1974, Vol. 1, p. 401.

[22]Ram Narayan Sinha, *Bihar Tenantry (1783-1833)*, Bombay, People's Publishing House, 1968, pp. 94-96; Narendra Krishna Sinha, *The Economic History of Bengal*, Calcutta, Frima K.L. Makhopadhyay, 1962, Vol. 2, pp. 169-73.

[23]Karuna Mukerji, *Land Reforms*, Calcutta, H. Chatterjee & Co., Ltd, 1952, p. 51.

[24]B.R. Mishra, *Land Revenue Policy in the United Provinces*, Benares, Nand Kishore & Bros, 1942, p. 102.

[25]*Ibid.*, p. 113.

[26]*Ibid.*, p. 120.

[27]*Ibid.*, pp. 157-59; Jagdish Raj, *The Mutiny and British Land Policy in North India, 1856-68*, Bombay, Asia Publishing House, 1965, pp. 164-68.

[28]In 1897, for instance, a *jimidar* from Kanchanpur in the Naya Muluk region claimed that he had reclaimed 1,200 *bighas* of virgin lands with settlers procured from India. "Order regarding Registration of Newly-Reclaimed Lands in Kanchanpur," Jestha Badi 3, 1944 (May 1897), *RRC*, Vol. 61, pp. 5-20.

[29]Henry Ambrose Oldfield, *Sketches from Nipal* (Reprint of 1880 ed.), Delhi, Cosmo Publications, 1974, p. 57.

[30]Statistics compiled on the basis of the revenue and expenditure figures of the government of Nepal for the relevant years.

[31]"Survey Regulations for Banke District," Chaitra Sudi 5, 1924 (March 1867), *RRC*, Vol. 37, pp. 267-87.

[32]"Commander-in-Chief General Dhir Shumshere's Order to Mukhiya-Jimmawal Jamadar Baliram Khatri Chhetri regarding Fiscal Concessions on Newly-Reclaimed Lands in Salyan," Bhadra Badi 10, 1940 (August 1883), in Yogi Naraharinath (ed.), *Sandhipatrasangraha* (A collection of treaties and documents), Dang, the editor, 2022 (1966), pp. 429-30.

[33]"Order regarding Cultivation of Winter Crops in Gulmi and Elsewhere," Marga Badi 3, 1923 (November 1866), *RRC*, Vol. 63, p. 350.

[34]An official notification published in 1921 took note of the fact that

"winter crops are not being cultivated in the hill areas because of the practice of letting cattle loose in the fields after the monsoon crops are harvested." The notification, therefore, directed that "in the future, cows, buffaloes, yaks, sheep, goats, pigs, asses, etc., shall be let loose only if attended by herdsmen, and grazed only on lands where crops have not been sown." *Gorkhapatra*, Ashadh 29, 1988 (July 13, 1921); *Regmi Research Series*, Year 3, No. 4, April 1, 1971, p. 93. Available evidence shows that these measures were as ineffective as their predecessors in promoting the cultivation of winter crops in the hill regions.

[35]John W. Mellor, *The Economics of Agricultural Development* (An Adaptation), Bombay, Vakils, Feffer and Simons Private Ltd., 1966, p. 19.

10

Thatched Huts and Stucco Palaces

We have now come to the end of our study of peasant and land-lord in nineteenth century Nepal. In the foregoing chapters, we have surveyed political developments in Nepal during the nine-teenth century and the nature of the Rana political system. Against the background of that survey, we have identified the major groups that were entitled by virtue of their political or economic status to appropriate the economic surplus generated by the Nepali peasant: *rajas*, *birta*-owners and *jagirdars*, *mukhiyas* and *jimidars*, and village moneylenders. We have also described the form in which the economic surplus was appropriated by these groups: taxes in money or in commodities, compulsory labour obligations or cash levies imposed in lieu of such obligations, and interest and other payments.

The study showed how the different categories of elite groups extracted surpluses from the peasantry through rents and taxes in the form of money, commodities, and labour. The majority of the population lived just on the level of subsistence, but produced enough to maintain a relatively affluent and high-living aristocracy and bureaucracy. An increase in the numbers of the aristocracy and the bureaucracy had necessarily to be balanced through an increase in rents, taxes, and labour services. Taken separately, it may be difficult to find fault with any individual tax or imposition. It is only the cumulative burden of all these taxes and impositions that left the peasant with a bare minimum of subsistence. The collection of additional exactions and servitudes in various forms, both illegal and extra-legal, by various layers of the aristocratic and bureaucratic hierarchy, as well as by the village elite at the local level, added considerably to that burden. Moreover, measures regarding the extraction of the agricultural surplus were applied during a period when agricultural productivity was virtually stagnant. If the state, or the landowning elite, or any of their local agents and functionaries, increased their share of the surplus, the peasant's income dwindled proportionately. Indeed, "without increases in productivity, squeezing a surplus from agriculture becomes a process of peasant exploitation—either by powerful landowners or by the state."[1]

The study showed that Nepal's political and economic system during the nineteenth century might aptly be described as an agrarian bureaucracy, or a system that depends upon a central authority for extracting the economic surplus from the peasantry. In essence, the system represented a coalition between the landowning and local elites ranged against the peasant. The entire state apparatus, and its legal and administrative policies, were geared to the task of extracting economic surpluses from the peasantry for the benefit of these groups. This institutional link between these two elite groups was one of the main factors responsible for the stability of the Rana political system for over a century.

Because the political elite formed the dominant segment of the landonwing elite, the agrarian policy of the state was determined by the *birta* and *jagir* owning members of the aristocracy and the bureaucracy. They had a vested interest in a high level of agricultural taxation, without any care for developing agriculture or improving the condition of the peasantry. Changes in the system were motivated primarily by their desire to facilitate the collection of rents and taxes. They had otherwise little interest in land and the peasant. For example, during his visit to England in 1850, Prime Minister Jung Bahadur seems to have been interested by an arsenal more than by anything else in London.[2] Nowhere do we find any evidence that he was interested in the industrial and agricultural progress that England had made during the nineteenth century. The class he belonged to had nothing in common with producers, nor any desire for change.

Indeed, official policy toward the peasant remained one of apathy and exploitation throughout the nineteenth century. Both the Ranas and their predecessors faithfully adhered to King Prithvi

Narayan Shah's dictum that "the King's storehouse is the people."[3] The peasant's worth consisted in his role as tiller of the land and a source of taxes and labour services. The official world of Kathmandu was shaken from its apathy only when the peasant found the burden so intolerable that he left cultivating the land.

Because agricultural productivity was low, and the burden of rents and taxes high, the majority of peasants in nineteenth-century Nepal lived on the margin between subsistence and destitution. The scale was tilted in favour of the latter condition when crops, always a gamble in the rains in the absence of organized irrigation facilities, failed because of drought. Famine was the inevitable result. Such a famine occurred in almost all parts of Nepal during 1863-66. The evidence contained in official orders and regulations promulgated from Kathmandu[4] shows that many people died of starvation in both the eastern and western parts of the Tarai region. The government sanctioned funds for relief, and even arranged for the supply of free food to destitute persons. However, these measures were shackled by bureaucratic procedures, and Kathmandu was forced to admit ruefully that "people are still dying of starvation." Its concern apparently stemmed from the realization that dead people can pay no taxes, and an attempt was made to force *jimidars* to bear the fiscal liability caused by deaths from famine. The concern felt by the landowning elite was apparently even less, because less than Rs 6,000 was collected in the course of a fund-raising campaign for the relief of the victims of the famine. The significance of the figure becomes obvious from the fact that *birta* incomes alone from the Tarai region during that period approximated Rs 900,000 a year.[5]

To be sure, the wider interests of the state occasionally acted as a constraint. We have seen how in nineteenth-century Nepal, as in other bureaucratic societies, the ruling elite's policies showed great preoccuptions with the social and economic status and activities of the peasantry.[6] The enactment of legislation seeking to impose limitations on the right of the landowning elite to evict their tenants may be cited as an example. Such examples would tend to show that the peasant gained whenever there was any conflict between the interests of the state and those of the landowning elite. Unfortunately, however, the two were too closely enmeshed to make such conflict a frequent occurrence. Indeed, their interests converged at points of vital importance to the peasant. The monetization of the rice-land tax system in the central hill region, for instance, was initially intended to facilitate the transfer of income from the land to the hands of *jagirdars*, but it also helped to increase the flow of cash revenue to the state treasury from newly-reclaimed and other lands that were under the direct control of the government.

The improvement in the peasant's status, if not in his income, during the third quarter of the nineteenth century, which was discussed in Chapter 8, was due less to the Ranas' concern for the peasant than to their desire to maintain their control over particular groups of landowning elite. One set of policies that the Ranas adopted with the objective of fulfilling this desire encouraged the peasantry in their attempts to free themselves from excessive dependence on the landowning bureaucracy. In *raikar* land, therefore, the goal of Rana policy was "to create and maintain an independent free peasantry with small holdings."[7] Official policy aimed at breaking the peasant-bureaucracy (*jagirdar*) nexus so as to prevent the bureaucracy from entrenching itself as a powerful and autonomous landed interest. It is noteworthy that the policy that the Ranas followed toward peasants on *birta* lands was radically different. The peasant-aristocracy (*birta*-owner) nexus was left by and large unchanged because the Ranas used their political authority to derive economic benefits for themselves primarily by virtue of their role as an important constituent of the traditional *birta*-owning aristocracy.

From the viewpoint of "the ratio between services rendered and the surplus taken from peasants,"[8] the relationship between *rajas*, *birta*-owners, and *jagirdars*, on the one hand, and the peasantry on the other, was clearly exploitative. These elite groups of ascriptive landowners were surperimpositions upon the local agrarian community. They performed no function that was an essential aspect of the peasant's way of life. Agricultural production would have remained unaffected, and productivity might even have increased, had these groups ceased to exercise political domination over the peasant. In fact, the landowning elite were allowed to exercise domination over the peasant only in order that the political elite might avoid an attack upon their political authority, while the village elite were necessary in order that this domination might be exercised effectively. In other words, the peasant bore the burden of sustaining not only his political and economic overlords, but also their local bailiffs.

Indeed, so exploitative was the agrarian system that the village moneylender's role in contrast appears relatively benign. To be sure, the moneylender was no more than a usurer, charging exorbitant rates of interest on the loans he supplied in order that a starving peasant might live from hand to mouth, but there is also an element of economic service in the role he fulfilled under the system. The moneylender, in fact, performed the essential function of credit supply in an agrarian society where traditionally it was the peasant who provided loans to his landlord rather than *vice-versa*. Among the groups that extracted the economic surplus generated by the Nepali peasant during the nineteenth century, the moneylender accordingly seems least exploitative. Nevertheless, this fact only highlights in bolder relief the unmitigated nature of the exploitation practised by the other groups.

The deleterious effect of such exploitation on the economic condition of the peasant might have been offset in some measure had the landowning elite invested at least a part of their income for raising the productivity of agriculture. There is, however, no evidence that they regarded land and the peasant as anything more than sources of income, which they used for unproductive investment and ostentatious consumption. For instance, construction of houses appears to have been one way in which the landowning elite of Kathmandu used their income. As Oldfield has recorded:

> Several of the sardars have during the last few years built large houses in different parts of the city. The sites on which they stand having been well selected, the ground levelled, and the surrounding buildings cleared away, give to them rather an imposing appearance, and make them contrast very strongly with the humble and dirty, but still very picturesque exteriors of the mass of the old Niwar dwellings in their neighbourhood.[9]

It may be argued that when the landowning elite used their income from rents in this manner, this would stimulate a demand for building materials and services of masons, carpenters and labourers, and thus foster economic activity in fields concerned with the supply of these materials and services. Nevertheless, this argument ignores two vital aspects of the question. Firstly, such a use of surplus agricultural production increases the wealth of the community, but not its stock of wealth-producing capital. Secondly, the temporary spurt of economic activity in the building construction industry is gained at the cost of keeping the peasant starved of resources which he might have been able to use for increasing agricultural production.

A sumptuous picture of the life-style of the landowning elite of Nepal emerges from contemporary descriptions of their palaces and reception rooms. Wright has given a graphic description of the residence of Prime Minister Jung Bahadur at Thapathali in Kathmandu, which consisted of "a succession of squares of gigantic houses, four or five stories in height," with "large, lofty, and ornamented" public rooms containing "pictures and carvings," and other "curiosities of Nepalese, Chinese and English manufacture, from a baby's frock to a reflecting telescope or an Erard's piano."[10] Nor was this affluence a monopoly of the Rana Prime Minister and his family. We have it on th e evidence of Oldfield that the public reception rooms in the houses of the lesser nobility also were "built in the English fashion, with lofty ceilings and glass windows, the walls of which are ornamented with mirrors and pictures, and the floors covered with Brussels carpets."[11] Oldfield also noted that these rooms were filled with "the most curious medley of useful and ornamental articles of English and French furniture. Steel fire-places, with marble mantlepieces; sofas, couches, easy chairs, billiard tables, and four-posted beds; candelabras, pianos, organs, glassware, vases, & c . . . ."[12]

A significant portion of the income that the landowning elite derived from rents was thus spent not on indigenous goods and services but on imported luxuries. It was because of their income from *birta* and *jagir* lands that "the Gorkhas are decidedly the best-dressed part of the population" in Kathmandu Valley.[13]

At the other end of the spectrum were the peasantry who tilled the land and paid the rents and taxes that were the source of the affluence of Kathmandu's elite. Hamilton has described the peasants of the eastern Tarai region as "extremely nasty, and apparently indigent."[14] He adds:

Their huts are small, dirty, and very ill calculated to keep out the cold winds of the winter season, for a great many of them have no other walls but a few reeds supported by sticks in a perpendicular direction. Their clothing consists of some cotton rags, neither bleached nor dyed, and which seem never to be washed. They are a small, hard-favoured people, and by no means fairer than the inhabitants of Bengal, who are comparatively in much better circumstances.[15]

Wright similarly noted that "the people are poor, and have few wants that are not supplied by their own country,"[16] and "can barely earn enough to feed and clothe themselves in the poorest manner."[17] There was, of course, a gap of nearly half a century between Hamilton's account of the reed-huts of the peasants and the stucco palaces of the Ranas in Kathmandu as described by Oldfield and Wright, but there is no evidence that the economic condition of the peasantry improved during the interval.

The contrast between the thatched huts of the peasantry, and the stucco palaces of *rajas*, *birta*-owners, and *jagirdars*, and the economic and psychological gulf that separated them, are accordingly a characteristic feature of nineteenth century Nepali economy and society. A substantial portion of the peasant's produce was reserved for individuals and groups who did not have any role in production, whereas his own income consisted of what was left after meeting their claims and exactions. As a consequence, whereas the landowning elite lived in ostentatious luxury, the peasant lived on the margin of subsistence, with neither the will nor the capacity to increase production.

To conclude: Under the agrarian system that existed in Nepal during the nineteenth century, resources were extracted from the peasant without any compensation, and neither the state nor the elite groups who absorbed these resources took serious interest in the lowly occupation of tilling the land and raising crops. Inasmuch as "productivity—or output per man-hour—depends largely, though by no means entirely, on the degree to such capital is employed in production"[18] such a situation had a deleterious impact on agricultural productivity. Low productivity due to inadequate capital investment in agriculture, the mainstay of Nepal's economy, was, consequently, the key link in the chain of historical causation that explains why Nepal remained a poor country during the nineteenth century.

NOTES

[1]Peter Dorner,*Land Reform and Economic]Development*, Middlesex, Penguin Books, 1972, p. 131.

[2]Pudma Jung Bahadur Rana, *Life of Maharaja Sir Jung Bahadur of Nepal* (Reprint of 1909 ed.), Kathmandu, Ratna Pustak Bhandar, 1974, p. 138.

[3]L.F. Stiller, S.J., *Prithvinarayan Shah in the Light of Dibya Upadesh*, Kathmandu, the author, 1968, p. 43.

[4]"Famine in Nepal," *Regmi Research Series*, Year 8, No. 11, November 1, 1967, pp. 201-208.

[5]See n. 44 in Chapter 3.

[6]S.N. Eisenstadt, *The Political System of Empires*, New York, Free Press, 1963, p. 208.

[7]*Ibid.*, p. 133.

[8]Barrington Moore, *Social Origins of Dictatorship and Democracy*, Middlesex, Penguin Books, 1969, p. 471.

[9]H.A. Oldfield, *Sketches from Nepal* (Reprint of 1880 ed.), Delhi, Cosmo Publications, 1974, p. 106.

[10]Daniel Wright (ed.), *History of Nepal* (Reprint of 1877 ed.) Kathmandu, Antiquated Book Publishers, 1972, pp. 13-18.

[11]Oldfield, *op. cit.*, p. 107.

[12]*Loc. cit.* Also read the description of Prime Minister Bhimsen Thapa's palace in Kathmandu by the poet Yadunath Pokhryal in Baburam Acharya, *Purana Kavi ra Kavita* (Old poets and poetry), Kathmandu, Sajha Prakashan, 2023 (1966), pp. 87-91.

[13]Wright, *op. cit.*, p. 28.

[14]Francis Hamilton, *An Account of the Kingdom of Nepal* (Reprint of 1819 ed.) New Delhi, Manjusri Publishing House, 1971, p. 169.

[15]*Loc. cit.*

[16]Wright, *op. cit.*, p. 69.

[17]*Ibid.*, p. 70.

[18]Ragnar Nurkse, *Problems of Capital Formation in Underdeveloped Countries*, Delhi, Oxford University Press, 1974, p. 9

Glossary

Adalat Goswara: Central Court of Justice, established in 1860

Adhiya: A system of sharecropping in the central region under which the land-lord (or the government) appropriated half of the produce as rent (or tax)

Adhiyar: Sharecropper in the eastern Tarai region

Amanat: A system of revenue collection through officials or functionaries appointed by the government or by *Jagirdars*

Anna: A unit of account equivalent to four *paisa*; 16 *annas* made one rupee, although there were local variations also

Asmani: Income from fines and penalties collected in the course of the dispensation of justice by the landowning or local elite

Bhith: Unirrigated agricultural lands and homesites in the Tarai region

Bigha: A unit of land measurement in the Tarai region. A *bigha* comprised 20 *katthas*, with a total area of 8,100 sq. yds. There were also local variations.

Birta: Land grants made by the king to individuals, usually on an inheritable and tax-exempt basis

Chardam-Theki: (1) A cash levy on rice-fields in the hill region. (2) A fee paid by a tenant to the landowning or local elites on the confirmation or renewal of his tenure.

Chaudhari: A functionary who collected land and other taxes in the *Parganna* in the Tarai region

Chuni: (1) Peasants who had not been enrolled under the *rakam* system. (2) Landholders in the far-western hill districts and the Tarai region who were listed as taxpayers in the official records.

Dara: A revenue sub-division comprising a number of villages in the Baisi region

Dhanahar: Irrigated lands in the Tarai regions where rice can be grown

Dharma-Kachahari: A high-level anti-corruption court established in 1870 and abolished eight years later.

Dhokre: Brokers who purchased *tirjas* from *jagirdars* and collected rents on *jagir* lands

Dware: A revenue collection functionary appointed by the *jagirdar* on *jagir* villages in the hill region

Ghiukhane: A cash levy paid on rice-fields in the hill region
Guthi: An endowment of land or other property for any religious or philanthropic purpose

Hulak: Porterage service for the transportation of mail or goods

Ijara: A revenue-collection contract
Ijaradar: Contractor; the holder of an *ijara*

Jagir: Lands assigned to government employees and functionaries in lieu of their emoluments
Jagirdar: The holder of a *jagir* land assignment
Jaisi: A caste comprising the offspring of an informal alliance between a Brahman man and a Brahman girl, married woman, or widow
Jhara: Compulsory and unpaid labour obligations due to the government
Jimidar: A functionary responsible for revenue collection in the villages of the Tarai region
Jimidari: Pertaining to a *jimidar*
Jimmawal: A functionary in the hill districts who allotted rice-lands to local peasants and helped *jagirdars* to collect rents
Jirayat: A plot of tax able land attached to a *jimidari* holding as part of the *jimidar's* emoluments

Kagate-Hulak: Porterage service for the transportation of mail. (See *Hulak*)
Khet: Irrigated lands in the hill regions on which rice can be grown
Kipat: A system of communal landownership prevalent among the Limbus and other Mongoloid communities in the hill region
Kut: A system of sharecropping in the central hill region under which the landlord (or the government) appropriated a specific quantity of the produce, or a stated sum in cash as rent (or tax)

Lokabhar: A system under which the local community undertook liability for payment of the stipulated revenue through a representative designated for that purpose

Mohi-Boti: Tenant's share of the crop
Mouja: A village as the primary unit of revenue administration in the Tarai region
Mukhia: A village headman in the hill districts.
Mukhiyabhar: A system under which the village *mukhiya* undertook liability in his personal capacity to pay the higher amount of revenue stipulated by a prospective *ijaradar*

Muri: (1) A unit of land measurement in the hill region. The area varied according to the grade: 1190 sq. ft. for lands of the best grade, and 1339 sq. ft., 1487 sq. ft., and 1785 sq. ft. for lands of inferior grades. Four *muris* of land were equivalent to one *ropani.* (2) A volumetric unit equivalent to 48.77 kg. of paddy, 68.05 kg. of wheat or maize, or 65.78 kg. of millet; one *muri* consists of 20 *pathis.*

Paisa: A unit of account. Four *paisa* made one *anna*

Pakho: Unirrigated high land or hillside land in the hill regions, on which only dry crops such as dry rice, maize and millet can be grown

Panchakhat: A term used to denote the crimes of bribery, smuggling, murder (including infanticide), assault resulting in the shedding of blood, and cow slaughter. These crimes were punishable through confiscation of property, banishment or degradation of the caste status, amputation, or death

Panchasala-Thek: A system of revenue-collection used in the eastern Tarai region between 1828 and 1849 under which local *chaudharis* stipulated the amount of revenue payable each year during a five-year period

Parganna: A revenue subdivision in the eastern Tarai region, comprising a number of villages

Pathi: A volumetric unit equivalent to 2.48 kg. of paddy, 3.40 kg. of wheat or maize, or 3.28 kg. of millet; 20 *pathis* make one *muri*

Raibandi: A system of periodic rice-land redistribution in the villages of the central hill region

Raikar: State-owned, taxable lands, which could be granted as *birta*, or assigned as *jagir*

Raja: The chief of a vassal principality in the western hill region

Rajya: A vassal principality in the western hill region

Rakam: Unpaid and compulsory labour services due to the government from peasants cultivating *raikar* (including *jagir*), *kipat* and *guthi* lands

Rekh: Proprietary landholdings in Jumla and elsewhere in the Baisi region

Ropani: A unit of land measurement used in Kathmandu Valley; comprising four *muris* of land. The actual area varied according to the grade. (See *muri*)

Sardar: A top-ranking officer in charge of military affairs

Saunefagu: A tax levied on the roof in the hill region

Serma: Homestead tax in the hill region

Talsing-Boti: Rent, or the landlord's share of the crop

Thaple-hulak: Porterage service for the transportation of goods. (See *Hulak*)

Thekbandi: A settlement with *mukhiyas* in their individual capacity for the collection of homestead revenue in the villages of the central hill region

Thekthiti: A settlement with the village community represented by the *mukhiya* for the collection of taxes on rice-lands and homesteads in the villages of the fareastern hill region and the Baisi region

Tirja: A letter of authority issued to a *jagirdar* entitling him to collect rents of his *jagir* lands

Upadhyaya: A pure Brahman, as distinguished from a Jaisi

Zamindar: A class of landowners in the eastern Tarai region and the Baisi region who were responsible for the collection of revenue from peasants living in the villages under their jurisdiction

Bibliography and Source Materials

The archival records in the possession of the Lagat Phant (Records Office) of the Department of Land Revenue in the Finance Ministry of His Majesty's Government provided the basic source materials for this study. Such materials were obtained also from the Ministry of Foreign Affairs, the Ministry of Law and Justice, and offices under the Guthi Corporation. Materials obtained from these sources are not individually listed below, but full references to all documents actually used have been given at appropriate places in the footnotes.

BOOKS AND ARTICLES

Nepali

Acharya, Baburam, *Purana Kavi ra Kavita* (Old Poets and Poetry), Kathmandu, Sajha Prakashan, 2023 (1966).
————,*Shri 5 Badamaharajadhiraja Prithvinarayan Shah* (The Great King Prithvi Narayan Shah), Kathmandu, His Majesty's Press Secretariat, Royal Palace, 2024-26 (1967-69), 4 pts.

Dikshit, Kamal, *Battis Salko Rojanamacha* (Jung Bahadurko) (Diary of Prime Minister Jung Bahadur's tour of the western Tarai and of the visit of the Prince of Wales in 1875), Lalitpur, Jagadamba Prakashan, 2023 (1966).

Government of Nepal, *Muluki Ain* (Legal Code). Various editions published between 1854 and 1870, including: Ministry of Law and Justice, *Shri 5 Surendra Bikram Shahdevaka Shasankalama Baneko Muluki Ain* (Legal code enacted during the reign of King Surendra Bikram Shah Dev), Kathmandu, the Ministry, 2022 (1965).

Naraharinath, Yogi (ed.), *Itihas Prakash* (Light on History), 4 vols., Kathmandu, Itihasa Prakasha Mandala, 2012-13 (1955-56).
— —,*Sandhipatra Sangraha* (A Collection of Treaties and Documents), Dang, the editor, 2022 (1965).

————and Acharya, Baburam, *Rashtrapita Shri 5 Bada Maharaja Prithvi Narayan Shah Devako Dibya Upadesh* (Divine counsel of the Great King Prithvi Narayan Shah Dev, father of the nation), (2d. rev. ed.), Kathmandu, Prithvi Jauanti Samaroha Samiti, 2010 (1953).

Nepali, Chittaranjan, *Janeral Bhimasena Thapa ra Tatkalin Nepal* (General Bhimsen Thapa and Contemporary Nepal), Kathmandu, Nepal Samskritik Sangh, 2013 (1956).

————,*Shri 5 Rana Bahadur Shah* (King Rana Bahadur Shah), Kathmandu, Mrs. Mary Rajbhandari, 2020 (1963).

Pokhrel, Balakrishna, *Panch Saya Varsha* (Five Hundred Years of Nepali Prose Literature), Lalitpur; Jagadamba Prakashan, 2020 (1963).

English

Baden Powell, B.H., *The Land Systems of British India*, (Reprint of 1892 ed.), 3 vols., Delhi, Oriental Publishers, n.d.

Bell, Clive, "Ideology and Economic Interests in Indian Land Reform" in David Lehmann, *Agrarian Reform and Agrarian Reformism*, London, Faber and Faber, 1974.

Bloch, Marc, *Feudal Society*, (Reprint of 1962 ed.), 2 vols., London, Routledge & Kegan Paul Ltd., 1975.

Bottomore, T.B., *Elites and Society*, (Reprint of 1964 ed.), Middlesex, Penguin Books, 1971.

Caplan, Lionel, *Land and Social Change in East Nepal*, Berkeley and Los Angeles, University of California Press, 1970.

Carr, E.H., *What is History*, (Reprint of 1961 ed.), Middlesex, Penguin Books, 1975.

Cavanagh, Orfeur, *Rough Notes on the State of Nepal*, Calcutta, W. Palmer, 1851.

Cipolla, Carlo M·, *The Fontana Economic History of Europe, Part II: The Emergence of Industrial Societies*, (Reprint of 1973 ed.), Glasgow, William Collins Sons & Co., Ltd., 1975.

Dobb, Maurice, *Studies in the Development of Capitalism*, (Reprint of 1946 ed.), London, Routledge & Kegan Paul Ltd, 1959.

Dorner, Peter, *Land Reform and Economic Development*, Middlesex, Penguin Books, 1972.

Eicher, Carl K., and Lawrence W. Witt, *Agriculture in Economic Development*, (Reprint of 1964 ed.), Bombay, Vora & Co., 1970.

Eisenstadt, S.N., *The Political System of Empires*, New York, Free Press, 1963.

Gadgil, D.R., *The Industrial Evolution of India in Recent Times, 1860-1939*, (Fifth ed.), Bombay, Oxford University Press, 1971.

Habib, Irfan, *The Agrarian System of Mughal India*, Bombay, Asia Publishing House, 1963.

Hagen, Toni, *Nepal: The Kingdom in the Himalayas*, Berne, Kimmerley and Frey, 1961.

Hamilton, Francis (Buchanan), *An Account of the Kingdom of Nepal*, (Reprint of 1819 ed.), New Delhi, Manjusri Publishing House, 1971.

———, *An Account of the District of Purnea in 1809-10*, Patna, Bihar and Orissa Research Society, 1928.

Hicks, John, *A Theory of Economic History*, London, Oxford University Press, 1969.

Hodgson, Brian H., "Some Account of the Systems of Law and Police as recognized in the State of Nepal," *Journal of the Royal Asiatic Society of Great Britain and Ireland*, Vol. 1, 1834, pp. 258-79.

Joshi, Bhuwan Lal, and Leo E. Rose, *Democratic Innovations in Nepal; A Case Study of Political Acculturation*, Berkeley and Los Angeles, University of California Press, 1966.

Karan, Pradyumna P., *et al.*, *Nepal: A Physical and Cultural Geography*, Lexington, University of Kentucky Press, 1960.

Kirkpatrick, W., *An Account of the Kingdom of Nepaul*, (Reprint of 1811 ed.), New Delhi, Manjusri Publishing House, 1969.

Kumar, Satish, *Rana Polity in Nepal*, Bombay, Asia Publishing House, 1967.

Lewis, W. Arthur, *The Theory of Economic Growth*, (6th ed.), London, George Allen & Unwin Ltd, 1963.

Landon, Perceval, *Nepal*, (Reprint of 1928 ed.), Kathmandu, Ratna Pustak Bhandar, 1976.

Lehmann, David, (ed.), *Agrarian Reform & Agrarian Reformism*, London, Faber and Faber, 1974.

Mcdougal, Charles, *Village and Household Economy in Far-Western Nepal*, Kirtipur, Tribhuwan University, n.d., (1968).

Madge, John, *The Tools of Social Science*, (Reprint of 1953 ed.), London, Longmans Green & Co. Ltd., 1968.

Marx, Karl, *Capital*, 3 vols, Moscow, Progress Publishers, 1965-68.

Mellor, John W., *The Economics of Agricultural Development*, (An Adaptation), Bombay, Vakils, Feffer and Simons Private Ltd, 1966.

Mishra, B.R , *Land Revenue Policy in the United Provinces*, Benares, Nand Kishore & Bros., 1942.

Moore, Barrington, *Social Origins of Dictatorship and Democracy*, (Reprint of 1966 ed.), Middlesex, Penguin Books, 1969.

Morrall, John B., *The Medieval Imprint*, (Reprint of 1967 ed.), Middlesex, Penguin Books, 1970.

Mukerji, Karuna, *Land Reforms*, Calcutta, H. Chatterjee & Co., Ltd, 1952.

Nadal, Jordi, "The Failure of the Industrial Revolution in Spain 1830-1914," in Carlo M. Cipolla, *The Fontana Economic History of Europe, Part II: The*

Emergence of Industrial Societies, Glasgow, William Collins Sons & Co., Ltd, 1975.

Nurkse, Ragnar, *Problems of Capital Formation in Underdeveloped Countries*, (Reprint of 1953 ed.), Delhi, Oxford University Press, 1974.

Oldfield, Henry Ambrose, *Sketches from Nipal*, (Reprint of 1880 ed.), 2 vols., Delhi, Cosmo Publications, 1974

Parsons, Kenneth H., "The Tenure of Farms, Motivation, and Productivity," in *Science, Technology and Development: Vol. III, Agriculture*, Washington, U.S. Government Printing Office, n.d.
————, "Agrarian Reform Policy as a Field of Research," in *Agrarian Reform and Economic Growth in Developing Countries*, Washington, U.S. Department of Agriculture, 1962.
Postan, M.M., *The Medieval Economy & Society*, Reprint of 1972 ed.), Middlesex, Penguin Books, 1975.

Raj, Jagadish, *The Mutiny and British Land Policy in North India, 1856-68*, Bombay, Asia Publishing House, 1965.
Ramakant, *Indo-Nepalese, 1816 to 1877*, New Delhi, S. Chand, 1968.
Rana, Maharaja Chandra Shum Shere Jung Bahadur, *Appeal to the People of Nepal for the Emancipation of Slaves and Abolition of Slavery in the Country*, Kathmandu, Suba Rama Mani, A.D., 1925.
Rana, Pudma Jang Bahadur, *Life of Maharaja Sir Jung Bahadur of Nepal*, (Re-Print of 1909 ed.), Kathmandu, Ratna Pustak Bhandar, 1974.
Regmi, Mahesh C., *Land Tenure and Taxation in Nepal*, 4 vols., Berkeley and Los Angeles, University of California Press, 1963-68.
————, *A Study in Nepali Economic History, 1768-1846*, New Delhi, Manjusri Publishing House, 1971.
————, "Preliminary Notes on the Nature of Rana Law and Government," *Contributions to Nepalese Studies*, Vol. 2., No. 2, June 1975, pp. 103-15.
————, *Landownership in Nepal*, Berkeley, Los Angeles and London, University of California Press, 1976.

Shanin, Teodor, *Peasants and Peasant Societies*, Middlesex, Penguin Books, 1971.
————,"Peasantry as a Political Factor," in Tendor Shanin, *Peasants and Peasant Societies*, Middlesex, Penguin Books, 1971.
Singh, L.R., *The Tarai Region of V.P.*, Allahabad, Ram Narain Lal Beni Prasad, 1965.
Sinha, Narendra Krishna, *The Economic History of Bengal*, 2 vols., Calcutta, Firma K.L. Mukhopadhyay, 1962.
Sinha, Ram Narayan, *Bihar Tenantry (1783-1833)*, Bombay, People's Publishing House, 1968.
Snellgrove, David, *Himalayan Pilgrimage*, Oxford, Bruno Cassirer, 1961.
Stiller, L.F., S.J., *Prithvinarayan Shah in the Light of Dibya Apadesh*, Kathmandu the author, 1968.

Temple, Richard, *Journals Kept in Hyderabad, Kashmir, Sikkim and Nepal*, 2 vols., London, W.H. Allen, 1887, vol. 2.

Thorner, Daniel, *The Agrarian Prospect in India*, Delhi, University Press, 1964.

————, "Peasant Economy as a Category in Economic History" in Teodor Shanin, *Peasants and Peasant Societies*, Middlesex, Penguin Books, 1971.

Warriner, Doreen, "Land Reform and Economic Development" in Carl K. Eicher and Lawrence W. Witt, *Agriculture in Economic Development*, Bombay, Vora & Co, 1970.

Wright, Daniel (ed.), *History of Nepal*, (Reprint of 1877 ed.), Kathmandu, Nepal Antiquated Book Publishers, 1972.

UNPUBLISHED MATERIALS

English

Department of Industrial and Commercial Intelligence. "Audyogik Survey Report" (Industrial survey reports for different districts) (English translations, mimeographed), Kathmandu, the Department, 2003-6 (1946-49).

NEWSPAPERS AND PERIODICALS

Nepali

Gorkhapatra

English

Journal of the Royal Asiatic Society of Great Britain and Ireland
Regmi Research Series

Index